POLITICAL VICTORY

POLITICAL VICTORY

BRIAN CROZIER

THE ELUSIVE PRIZE
OF MILITARY WARS

TRANSACTION PUBLISHERS
NEW BRUNSWICK (U.S.A.) AND LONDON (U.K.)

Library of Congress Catalog Number: 2005050597
ISBN: 0-7658-0290-2
Printed in the United States of America

Library of Congress Cataloging-in-Publication Data

Crozier, Brian.
    Political victory : the elusive prize of military wars / Brian Crozier.
        p. cm.
    Includes bibliographical references and index.
    ISBN 0-7658-0290-2 (alk. paper)
        1. War—Public opinion. 2. Politics and war. 3. Civil-military relations.
    4. Influence (Psychology) I. Title.

U21.2.C76 2005
303.6'6—dc22   2005050597

Contents

Part 6: Looming Threats

Preface

Success in war has always been difficult to measure. What is judged successful by the military leaders may not be so judged by the political leadership (often, though not invariably, those who took the decision to launch the war in the first place, whether in an aggressive or defensive sense). Nor will the wider public, at least in a Western-style democracy, necessarily agree with the self-congratulatory moods of the victorious military, or of the political leaders who sanctioned or approved the decision to fight whichever country or regime made war seem necessary or even inescapable.

Total defeat and its counterpart, total victory, tend to be judged in military terms, even by the politicians in power; whereas the voting public (again, in a democracy), while usually ready to greet a military victory with applause, instinctively reserves the right, when the applause has died down, to ask the question: Was it really worth it?

The main purpose of this book is to look at a selection of wars involving democracies in the twentieth century and the early twenty-first, and judge them, mainly, not by a military victory or defeat but by the success or failure of the political outcome.

By far the most important wars of the last century came in its early decades: World Wars I and II. In both cases, the main players were the same: on the aggressive side, Germany; on the defensive side, three key democracies: the United States, the United Kingdom, and the French Republic. In each case, the German aggressor was defeated. Having won the First World War, the Western democracies lost the peace; having won the second, they won the peace as well. In each case, another major player was Russia: Tsarist Russia in World War I, Communist Russia in World War II. In the second one, there was another major participant: imperial Japan. On the German side, the legacy of defeat in World War I was the advent to power of Hitler and the Nazis. On the Russian side, the defeat of the tsarist forces paved the way for Lenin's Communist revolution.

The total defeat of the Nazi regime in 1945 left the Western allies in charge of about two-thirds of Germany's population, thus enabling the victors to convert the vanquished to democracy. This was a major, epoch-making political victory; although, unfortunately, East Germany, under Soviet control, came under Communist rule. Defeated Japan, however, under the victorious American General Douglas MacArthur, also became a democracy.

After World War II, a major problem facing Western Europe's two leading democracies–Britain and France–was the future (if any) of their respective colonial empires. Although both countries emerged from the war on the winning side, the contrast between them was deep. For geographical and other reasons, France was defeated by Germany in the first phase of the war and its government, under Marshal Pétain, collaborated with the Nazis. Partly because Britain is an island, it escaped German occupation, although its army had been defeated on French soil.

Its wartime leader, Winston Churchill, led his country to victory after the humiliating retreat from Normandy in 1940. Having done so, however, Churchill was ousted in the first postwar elections and the Labour Party, led by Clement Attlee, came to power. Churchill had famously declared that he had not become prime minister to preside over the dissolution of the British Empire; Attlee made it clear that he aimed at independence for Britain's former colonies. The result was that Britain avoided full-scale wars in the inevitable postwar colonial conflicts.

The same was not true of France, particularly in its two major colonial areas: Southeast Asia and North Africa. The French had emerged from the war defeated and demoralized. Their wartime leader-in-exile, General Charles de Gaulle, had spent most of the war in Britain. Although he returned in triumph when the German occupying force surrendered in Paris on 23 and 24 August 1944, his position was far from secure. Almost exactly a year later, he faced a totally unpredictable crisis in Indochina, one of France's main colonies. First, the Japanese occupying force seized control of Vietnam on 9 March 1945, and installed the semi-captive emperor, Bao Dai, as the puppet head of state. When Japanese resistance collapsed a few months later, after the atom bombing of Hiroshima on 6 August, the Moscow-trained Vietnamese Communist leader, Ho Chi Minh, proclaimed his country's independence, with American approval.

Full independence was far from what de Gaulle had in mind. Already, on 24 March that year, he had offered the Indochinese peoples "freedom" and "economic autonomy," but within the French Union.

This would have amounted, at best, to a limited autonomy, with France in ultimate control. The war that followed lasted nine years and ended with the total defeat of the French forces in the northern "fortress" of Dien Bien Phu. The outcome, at the Geneva conference that followed, was the division of Vietnam into two entities: the communist north, supported by the Soviet Union and communist China; and the south by the United States and Great Britain.

Back to Vietnam. America's political and military involvement in this troubled region, which followed France's withdrawal, had nothing to do with colonialism, but has to be seen in the context of much wider issues: the Cold War and the mounting challenge of international communism to the Western democracies. During the second Vietnam War, the U.S. role was initially confined to military assistance and supplies to the South Vietnamese forces. From 1965 on, however, the United States was directly involved in a war that was doomed to failure on the U.S. side. At the end of March 1973, the last U.S. troops left Vietnam, and South Vietnam surrendered on 30 April. The reasons for America's defeat are analyzed in detail in the relevant chapter.

Within weeks of France's military and political defeat in Vietnam, a parallel conflict broke out much closer to the colonial nation: in Algeria, across the Mediterranean from southern France. There had been signs of colonial restiveness as early as May 1945, when Algerian nationalists clashed with the French forces. On that occasion, more than 1,000 Algerians had died, as against only eighty-eight French.

On 1 September 1947, all Algerians were granted full French citizenship, with an Algerian assembly of their own and financial autonomy. Such concessions were of the "feel good" category, but did not begin to meet Algeria's nationalist aspirations. Open revolt against French colonial rule was bound to come, and did, on 31 October 1954. The rebels had given themselves a name: the Front de Libération Nationale or FLN. Nearly eight years later, on 3 July 1962, France formally proclaimed Algeria's independence.

The nearest Britain came to a colonial war comparable to the French wars in Vietnam and Algeria was the Malayan Emergency (as it was known). The challenge to the colonial authority came from the Malayan Communist Party (MCP), providing a parallel to the anti-French colonial war in Vietnam (but not in Algeria), Again, as in Vietnam, the local Communists enjoyed support from the Chinese CP. The

Malayan Emergency never developed into a full-scale war, but went on for twelve years, ending on 12 July 1960 with the total defeat of the guerrillas. Politically, two objectives had been achieved: Malaya had been delivered from the threat of Chinese communism; and on 31 August 1957, Britain–the colonial power–conferred independence on Malaysia (Malaya, plus the states of Sabah and Sarawak on the island of Borneo). The main beneficiaries were the Malay majority, through their governing party, UMNO (the United Malays National Organization). To a lesser degree, the main minorities–the Chinese (through their Malayan Chinese Association or MCA), and the Indians (through the Malayan Indian Congress or MIC)–were also beneficiaries of Britain's victory.

From Churchill's retirement, spent mainly in writing his important memoirs, Britain's wartime leader may have wondered how he would have handled the postwar crisis sparked by India's demand for independence in 1947. A complicating factor arose when the peaceful handover envisaged on both sides collided with separatist calls from India's Islamic leader Mohammed Ali Jinnah, The independent Muslim State Jinnah insisted upon would have to be carved out of Indian territory.

Nor was this the only complicating factor. An apparently insoluble problem was presented by the mainly Muslim state of Kashmir. As it happened, the ancestors of India's leader, Jawaharlal Nehru, were Kashmiris, which led him (mistakenly to outside observers) to insist on including Kashmir in an independent India. This demand was of course unacceptable to Jinnah. The delayed outcome was the India-Pakistan war of 1971-72, and continuing tensions between the two post-colonial states, culminating in nuclear confrontation (fortunately short of war) in the late 1990s.

An encouraging end to a major conflict, though long delayed, marked the Soviet Union's attempt to subjugate Afghanistan. After years of efforts to gain political control over this complex and mountainous country, the Soviet Union did establish a satellite government in Kabul, but faced challenges from various tribes, at war with each other as well as with the ideological colonialists

After fourteen years of fighting, the Mujaheddin guerrillas toppled the Soviet-controlled central government in April 1992. Three years earlier, in 1989. the Soviet Union's last leader, Mikhail Gorbachev, had given up the struggle and withdrawn the invading force from Afghanistan. The withdrawal sparked off a series of anti-Soviet demonstrations or uprisings in various republics of the USSR, culminating in the collapse of the Soviet system.

One year before Malaya's independence, Britain and France had been involved, as allies, in a spectacular military venture, known nowadays as the Suez fiasco. On 26 July 1956, the Egyptian leader, Gamal Abdel Nasser, nationalized the Suez Canal Company, thus asserting his country's sovereignty over the famous Canal predominantly controlled by France and Britain. At that time Britain's prime minister was Sir Anthony Eden, best known for his long service as foreign secretary. As Eden revealed in his later memoirs, he regarded Nasser as an incipient Arab Hitler. Against strong American opposition, Eden and the French premier, Guy Mollet, decided to invade Egypt, and, moreover, to encourage Israel to do likewise.

The whole operation did indeed end in fiasco. Sick and demoralized, Eden resigned, and the calmer, wiser Harold Macmillan took over. In France, Guy Mollet remained in power, but only until May 1957. The "defeated" political victor was Nasser.

Eleven years after the Suez debacle, Nasser suffered an even more stunning defeat–at the hands not of a great power but of a tiny Middle East country with a mainly non-Arab population. I refer, of course to Israel, and to the Six-Day War of 1967. After Hitler's massacres of the Jewish population in Europe, victorious Britain pledged its support for Zionist efforts to turn Palestine into a home for world Jewry. Inevitably the ensuing partition of Palestine into Arab and Jewish zones caused mounting tensions between the rival communities, which broke into war in June 1967.

Although heavily outnumbered by the armies of the Arab states, the Israeli forces were highly trained and brilliantly commanded. They defeated the Arab forces in a mere six days. But the Israeli victory, far from bringing peace to Israel, exposed the new country to continuing hostility. Although defeated a second time, Colonel Nasser had reasons to rejoice.

Let us move on, by about a decade, to an unexpected little war involving Britain, which broke out in April 1982 when Argentina invaded and captured the Falkland Islands, 400 miles from its southern shores and 8,000 miles from their colonial power: Great Britain. Sovereignty over these islands, named Las Malvinas by the Spanish-speaking Argentinians, had long been contested, between France and Spain, in addition to Britain and Argentina, but had remained under British rule since 1833. Not irrelevantly, their "population" included not only 2,000 humans, but 600,000 sheep.

Britain's prime minister, Margaret Thatcher, recently elected, responded to the Argentinian occupation by sending a naval and military force to the waters of southern Argentina, which defeated the invaders and restored British sovereignty over the contested islands. It also restored the protection against possible foreign persecution by the invaders. Moreover, this short but effective confrontation significantly boosted the popularity and authority of Britain's first female prime minister. In other words, Margaret Thatcher could legitimately claim to have achieved an unusually impressive political victory by military means.

In more recent times, three wars sharpened the great divide between military and political success: the two Gulf Wars, in 1980 and 1990-91; and the Kosovo war of 1999.

The first Gulf War was a manifestation of dictatorial hubris, on the part of the Iraqi dictator Saddam Hussein. Having achieved absolute power by ruthless terror, he craved international prestige and sought to achieve it by invading neighboring Iran. The war lasted eight years and ended in a humiliating defeat for Iraq.

The second Gulf War, conducted during the "collapse" phase of the Soviet Union, brought well-deserved diplomatic credit to President George Bush of the United States for his remarkable skill in building a UN-backed coalition in support of his plan to teach a lesson to Iraq and its ruthless dictator, Saddam Hussein. Militarily, it was an admirable success and lifted the Iraqi threat to oil-rich Kuwait: the pretext of the war. Politically, it was a disastrous failure, in that it left Saddam in power.

This key factor was an issue between the Allied commander-in-chief, General H. Norman Schwarzkopf, who is understood to have favored removing the dictator; and, on the opposite side, President Bush himself and General Colin Powell, chairman of the Joint Chiefs of Staff. Bush and Powell prevailed. The message to the outside world was that the war had been won. Militarily, it had; but politically, the peace had been lost.

Unlike the Gulf War, NATO's offensive against Serbia was the first in the half-century of the Alliance to have been launched not only without UN approval but outside the terms of reference of the Atlantic Treaty. The rationale of the offensive was to end the Serbian oppression of the Kosovar Albanian minority. The concentration on the air offensive saved allied lives but left the Serbian forces essentially undefeated. Interestingly, the delayed outcome of the war was the defeat of the Serbian dictator President Slobodan Milosevic in democratic elections and the prospect that he would be brought to trial for war crimes: an unpredictable political victory for democracy.

Another war, which happened as long ago as 1950, may yet qualify for space under the heading of "Looming Threats." At the international conference in Geneva, already mentioned, a protracted attempt to settle another conflict of a somewhat different character ended in stalemate; I refer to the Korean war of 1950-1953 between the Communist North (supported by Moscow and Beijing) and the UN-supported South.

The United States and Britain, along with France and many other nations, had taken part in the Korean War of 1950-1953: the first war of its kind, a defensive war authorized and launched by the United Nations. In the final phase of the Pacific war, Stalin had declared war on Japan, belatedly, on 8 August 1945, only two days before Japan capitulated. On the 10th–the day of Japan's surrender– Soviet forces crossed the border into Korea and reached the 38th parallel (the agreed dividing line between Soviet and American forces) by 26 August, whereas the U.S. forces did not land in Korea until 8 September.

The Soviet Union established a North Korean regime under Stalin's protégé, Kim Il Sung; while south of the 38th parallel, the UN supervised a free election that brought Syngman Rhee to presidential power. Five years on, with Stalin's approval, Kim Il Sung's Communist forces crossed the 38th parallel on 25 June 1950. The Korean War had begun. On the international scene, however, Stalin had committed a major error by ordering the Soviet representatives to boycott the UN Security Council. Summoned by the UN's secretary-general, Trygve Lie, the Security Council met, and in Russia's absence, passed a resolution calling on member states to provide military aid to South Korea. Thus the Korean War began, rendered more complex by Communist China's decision to enter the conflict on North Korea's side.

The war ended in stalemate, leaving the tyrannical Communist regime in power in the North while democracy blossomed in the South. There can be no doubt that the Communist North remains a threat as clashes in recent years demonstrate.

How many times in this twenty-first century have democracies waged a war, and won it both militarily and politically? To expand on points already raised, more often than not, as this book demonstrates, military victory has been followed by a lost peace. It was too early, when these lines were written, in late 2002, to say whether the U.S.-led war on international terrorism will end in a lasting victory for the West, although the rapid overthrow of the barbarous and tyrannical Taliban regime in Afghanistan was encouraging. In all the wars considered, the determining criterion of strategic success should have been the attainment and maintenance of democratic systems, whether in Europe or in Asia, in Latin America, in the Middle East or in Oceania.

The militant exponents of aggressive dictatorships during World War II were Nazi Germany and imperial Japan. Both have turned into successful and non-aggressive countries, no longer a danger to their neighbors or to the wider world. The Soviet Union, no less anti-democratic than Nazi Germany, defeated the latter after Hitler committed his major strategic error of invading Stalin's territory. Forty-six years after the war ended, the totalist tyranny of the Soviet Union collapsed. It is of major importance to the West that Russia should move away from its long tradition of political autocracy

Japan was luckier than the Soviet Union: utterly defeated in 1945, and thanks to the architect of its defeat, General Douglas MacArthur, the Japanese opted—successfully—for democracy, and ceased to present a threat to East Asia. The most likely cause for another major war in Asia is the fact that China is still, at the start of the twenty-first century, under totalist rule, despite its encouraging switch from a failed Communist economy to a successful capitalist one. The cause of democratic Taiwan greatly exceeds in fundamental importance the West's interest in trading with the giant but still autocratic tyranny of Communist China.

A further look at this major issue from the viewpoint of looming threats, is considered in the epilogue.

Part 1

World-Scale Failure and Success

1

The First Two Franco-German Wars (1870-1914)

Three times in sixty-nine years–a human life–Germany and France were at war: in 1870, in 1914 and in 1939. The "peace" that followed the first of these wars led to the second one; and the peace of the second made the third inevitable.

The guilt for the first of these international clashes falls on Napoleon III, who evidently aspired to emulate the exploits of Napoleon I, the military genius who had occupied much of Europe, despite his major defeats in Russia, at Trafalgar, and at Waterloo. The blood relationship between the first and third Napoleons was minimal, however. Napoleon II was the son of the emperor. At birth, he had been given the title of king of Rome, which, under the contemporary rules of succession, was interpreted as heir to his father's empire. However, he died in 1832 of consumption, aged only twenty-one.

Napoleon III's relation to the emperor was more remote: he was the son of Napoleon the Great's third brother, in other words, a cousin of the ailing Napoleon II. On 2 December 1852, he proclaimed himself emperor under the logical title of Napoleon III. In reality, though not in legal terms, his was an absolute dictatorship, though not without its good aspects. Inspired by the French architect Baron Georges Haussmann, he ordered the rebuilding of Paris into the great capital admired by generations of visitors. He declared war on Italy in 1859, seizing Nice and the province of Savoie.

The First Franco-German War: 1870-71

Despite complex setbacks in Mexico and Poland, the emperor declared war on Prussia on 19 July 1870. This would be his nemesis. The climax came in September with the disastrous Battle of Sedan. The political victor was the great unifier of Germany, Prince Otto von Bismarck. Sedan marked the apotheosis of Bismarck and

3

the dishonorable end of Napoleon III, who was led in captivity to the Castle of Wilhelmshöhe before being allowed to retire in the Kentish town of Chislehurst, where he died three years later.

Facing famine, Paris surrendered to the Germans on 28 January 1871. In late March, a disparate group of Radicals, Moderate Republicans, and members of the Communist First International set up a municipal council calling itself the Commune, the first Communist attempt to seize power. It was a short-lived one. In April the National Assembly at Versailles sent a military force to Paris, and armed clashes followed, together with arbitrary executions on both sides.[1]

Meanwhile, Bismarck had created a new German empire, by negotiating separately with each German state, The new Reich, as it was known, included no fewer than twenty-five states: four kingdoms (Prussia, Bavaria, Saxony, and Württemberg); five grand duchies; thirteen duchies and principalities, and the three "Free Cities" of Hamburg, Bremen, and Lübeck. Bismarck had also annexed the long-disputed territory of Alsace-Lorraine, which was now designated as a *Reichsland* and made the common property of all the German states. Its new status was emphasized by the designation of *Statthalter* or imperial governor.[2]

The Second Franco-German War: 1914-18

On 18 March 1890, Bismarck resigned as chancellor of the German Empire. His departure was not voluntary: it was ordered by the new emperor, Wilhelm II, who had ascended the throne on the death of his father, Friedrich III. Much has been written about the charm and intelligence of the new emperor, but also on his headstrong and impulsive character.

Democracy of a kind flourished during his reign, but so did international pressures from Germany and a massive expansion of the German naval and military machine, along with a colonial policy of aggressive competition with Britain in Africa. In elections to the Reichstag (the German parliament) in 1912, the Socialists, backed by 4.25 million votes, had 110 deputies and dominated the legislature. Half way through the following year (1913) the Army and Finance bills were passed, providing for a peacetime increase of the Army, from 544,000 to 870,000 men. The cost of the increase was estimated at one billion marks. Despite assumptions or allegations to the contrary, the kaiser's government did not deliberately plan a second war on France in 1914.

Its main military preoccupation was with the necessity, as it was seen, of bringing tsarist Russia to heel. Then came the unpredictable assassination of the archduke of Austria, Franz Ferdinand, at Sarajevo, in Bosnia-Hercegovina, on 28 June 1914. For the kaiser and the German High Command, the question was whether or not to stand by Austria-Hungary. The Germans authorized Austria to act against Serbia and promised support, should Russia decide to intervene.

Clearly, they had not fully worked out the possible consequences. Their immediate expectation was that the hostile powers would simply give way, but they did not. The chief of the German General Staff, Helmuth von Moltke, warned Russia against mobilization, but the Russians went ahead anyway. In the tense days that followed, Moltke and his General Staff worked out a plan of action: they would immediately declare war on France, defeat the French, then take on the Russians. They calculated that their forces would overrun France in six weeks, and Russia in six months. The double calculation was wrong at both ends.

There is no need, in the context if this chapter, to deal in detail with the battles of the Great War, The relevant point is that Germany was defeated by the Franco-British alliance, with the support of the Americans, from 1917.

But things did not work out that simply. The German armies— seven of them—overran Belgium, as planned, but by the end of 1914 the clash of armies had taken on a new guise: a trench war of mutually hostile immobility. Meanwhile, on the Eastern front, the German armies were thrashing the Russians. By the end of 1916, the Austro-German forces had totally defeated the tsarist army. The following year the German High Command reached a secret deal with the self-exiled Russian Communist leader, Lenin. Their outlook was short-term: they would enable Lenin to return to Petrograd, so that he could overthrow the provisional government that had taken over from the defeated tsarists, and put an end to Russian resistance. In contrast, Lenin's own standpoint was long-term. Whatever the price, he would end the war and get on with his planned revolution.

The Germans brought Lenin and his handful of Bolshevik exiles from Switzerland to Russia in a sealed train and gave him 50 million marks in gold. His first attempt to seize power failed, but a second one succeeded later in that fateful year: 1917.[3] The Eastern Front was now clear.

In March 1918, the new Russian revolutionary government was forced to sign the Treaty of Brest-Litovsk: a total capitulation. The German beneficiaries gained control over 56 million citizens, 79 percent of Russia's iron and 89 percent of its coal.

On the Western front, Ludendorff, taking the kaiser's name in vain, launched what he called "the emperor's battle" (much against the emperor's wishes, incidentally) and scored minor and separate victories over the French and the British. In July and August respectively, the two countries struck back, effectively. In that decisive summer, new forces and tons of war equipment poured in from the United States. Eighth August 1918 became known as the "Black Day" for the German Army, as the Allied forces broke through the German defenses. In the first days of the Allied attack sixteen German divisions were wiped out. So decisive was the German retreat that soldiers as well as tanks and guns were left behind. On 14 August, Ludendorff advised the Kaiser that negotiations with the Allies would have to come soon. .A few weeks later, on 29 September, he was calling for peace at any price.[4]

By then, Germany's allies too were collapsing. In September, Bulgaria stopped fighting, and Austria-Hungary was on the verge of it. Alarmed, Ludendorff made peace overtures to the United States, whose president, Woodrow Wilson, had spelt out his terms in a list of fourteen points, which Ludendorff, unfortunately for him, did not read immediately.

For the sake of clarity in a complex situation, the list may be summarized as follows:

1.   Open covenants openly arrived at.
2.   Absolute freedom of navigation alike in peace and war, except as the seas might be closed by international action to enforce international covenants.
3.   The removal, so far as possible, of all economic barriers.
4.   Adequate guarantees that armaments would be reduced to the lowest point consistent with domestic safety.
5.   An impartial adjustment of all colonial claims on the principle that the interests of the population must have equal weight with the claims of the government.
6.   The evacuation of Russian territory and the free determination of her own political and national policy.
7.   Evacuation and restoration of Belgium.
8   Evacuation and restoration of French territory and righting of the wrong done to France in the matter of Alsace-Lorraine.

9.   Evacuation and restoration of Roumanian (sic), Serbian, and Montenegrin territory, together with access to the sea for Serbia.

12. The Turkish parts of the Ottoman Empire to be given a secure sovereignty, but the other nationalities to be given an opportunity for autonomous development, and the Dardanelles to be permanently opened to the ships of all nations under international guaranties (sic).

13. An independent Poland, to include territories indisputably Polish, with free and secure access to the sea.

14. A general association of nations to be formed to afford mutual guaranties of political independence and territorial integrity to great and small States alike.

When Ludendorff did read them, he was startled. As he understood them, the fourteen points spelled out recognition by Germany of its total defeat.

The most important message to the German High Command was simply that all territories seized or occupied by the German forces would have to be restored to the countries concerned. Unnerved by his reading, Ludendorff resolved to resume the war. He was overruled, however, and resigned on 26 October. The head of Germany's General Staff, Field Marshal Paul von Hindenburg, remained at his post.[5]

On 9 November, Wilhelm II abdicated and fled to the Netherlands. Karl Liebknecht, leader of a Communist group known as Spartakus, had been plotting to proclaim a Soviet republic, but was forestalled by a Social Democrat, Philipp Scheidemann. These were troubled times. There was street fighting in Berlin and two extreme leftists, Liebknecht and Rosa Luxemburg, were arrested and beaten to death.

On 19 January 1919, Germany's first united and democratic republic emerged. A newly elected National Assembly met at Weimar on 6 February. The territory which it would run soon became known as the Weimar Republic, In retrospect, it would probably have been in the general interest for the victorious Western Allies to discuss the terms of peace in detail with this new and democratic republic. Instead, the Allies decided on absolute intransigence. There would be no discussion on equal terms. The Allies had won the war, and would now impose the peace terms on the defeated enemy.

Germany would have to pull out its troops from Rumania, Austria-Hungary, Turkey and, when judged possible, from Russia. In material terms, the Germans would have to surrender 5,000 locomotives, the same number of lorries and 150,000 freight cars, plus

160 submarines and many other warships. Hostilities ceased on the Western front, at 11 A.M. on 11 November 1918; which became known as Armistice Day, and would be observed every year on that date and at that time by the victorious Allies.

An important date was 18 January 1920, when the peace conference met in Paris–without any German representatives.

During the war, the Western alliance had attracted the support of many minor states, so that at its conclusion no fewer than twenty-seven nations were represented at the Paris peace conference. The three leading ones were of course the United States, France, and Great Britain, respectively through President Woodrow Wilson, Prime Minister Georges Clemenceau, and Prime Minister David Lloyd George, all three of whom had interesting though contrasting personalities and careers.

Woodrow Wilson, who became the twenty-eighth president of the United States, was an idealist as well as a politician; indeed, it could be said that his idealism turned him into a statesman. It could also be said that his sternly Presbyterian father, a distinguished theologian and minister (whose own father had emigrated from Ulster), made and left a strong impression on the future president. Elected to the White House in 1912, Wilson lived up to his principles of never abusing the power of the United States, and always respecting the interests of weaker states.

When the Great War broke out on 4 August 1914, President Wilson opted for neutrality. By the spring of 1917, however, outraged by the continual sinking by German submarines, of Western liners often with many American passengers, he changed his mind. And on 6 April that year, the United States declared war on Germany.

U.S. forces joined with the French and British in their clashes with the German forces. On 4 October 1918, the German and Austrian governments appealed to President Wilson for an armistice, having accepted his fourteen points as a basis for peace.

No greater contrast could be imagined than between the personalities of Wilson and the French prime minister (Président du Conseil) Georges Clemenceau. Fifteen years older than Wilson, he obtained a doctorate in medicine in 1869 (although thereafter, he scarcely practiced). He went to the U.S. on a short study course. Back in France in 1870, he entered politics, was elected mayor of Montmartre (the "arts" quarter of Paris) and was soon elected deputy to the French Parliament, as leader of the extreme Left. Soon nicknamed the "tumbler of ministries" he spent most of his public time attacking other politicians.

Appointed president of the council (prime minister) in 1906 he created a Ministry of Labor but devoted most of his time and energies to the leadership of France during the war. Typically, his favorite phrase during the war years was: "*Je fais la guerre*" ("I'm fighting the war"). An authoritarian and uncompromising leader, widely known as "the tiger," he brought various pacifists to trial. His brutal, anti-clerical attitude made him many enemies. Nevertheless, his attitude changed when Germany was defeated: he supported a continuing policy of alliances and accepted the sacrifice of France's Rhine frontier in the interests of the Entente Cordiale. Beaten in the new presidential elections, in 1920, he retired in a long-lasting haughty mood.

The third Allied leader (though not necessarily in that order) was the fiery British politician David Lloyd George, who dominated British politics for sixteen years from 1906 to 1922. A Welshman by parentage, though born in Manchester, his father was headmaster of a school there and his mother was the daughter of a Baptist minister. As a young man, he became well known for his romantic good looks, and his marriage did not stop his long string of love affairs.

Elected to the House of Commons in1890, he quickly made a name for himself as a witty and spellbinding orator. During his early political years, perhaps surprisingly, he was notoriously a pacifist and bitterly opposed the Boer War in South Africa (1899-1902). On a visit to Germany in 1908, he became an enthusiastic convert to Bismarck's scheme of insurance benefits. Back in England, he introduced health and unemployment benefits on a similar basis under his National Insurance Act of 1911, which became the basis of Britain's "welfare state."

When the Great War broke out three years later, he went through a brief phase of pacifist isolationism, but soon accepted his appointment as minister of munitions and secretary of state for war. At the risk of shocking his Liberal colleagues and the civil servants, he imported experienced businessmen to build up the munitions industry while using his eloquence to convert organized labor to his methods. He became prime minister during the war, but not until December 1916. The general verdict of his wartime leadership is that he did his job with considerable vigor. He joined his French and American colleagues and allies in Paris representing the British Empire as well as Britain at the conference that signaled the aftermath of the First World War.

Wilson, greeted by vast and enthusiastic crowds in Paris, was–to the disappointment of the absent Germans–less interested in his own fourteen points (which had caused the Germans to cease fighting) than in his new idea: for a "League of Nations" designed to prevent any future wars.

Lloyd George, who represented not only Britain but the British Empire, was primarily interested in two proposals that had dominated a recent Parliamentary election: that Germany should be made to pay for the war, and that war criminals should be brought to justice. Clemenceau's primary interests were: revenge and provisions for the future security of his country.

Complicating factors were the Russian Revolution and the civil war that followed. The French and British were strongly opposed to the Russian Bolsheviks, who were not invited to the peace talks. There were many other points at issue between the Allies, notably the French and British insistence that Germany should be made responsible for the entire costs of the war, to which President Wilson objected. Finally, on 7 May, a draft treaty was presented to a German delegation, which had arrived in Paris a week earlier.

Not surprisingly, the head of the German delegation, Count Ulrich von Brockdorff-Rantzau, took vigorous exception to the terms of the draft. His argument was that these were out of harmony with the conditions that had caused Germany to lay down its arms. Moreover, he added, many of the clauses were simply impossible of fulfillment. The victorious Allies, however, were in no mood to make concessions.

The final draft of the Versailles Treaty included the following terms:

- Alsace-Lorraine to be returned to France;
- Upper Silesia and West Prussia to be ceded to Poland;
- North Schleswig to Denmark;
- Danzig to be a free, independent city;
- A ban on Germany's plan to unite with Austria;
- The banks of the Rhine to be permanently demilitarized;
- The coal-rich Saar to be administered by the League of Nations (not yet in existence at that time) for fifteen years, with the coalfields under direct French control;
- The German army to be limited to 100,000 men, and the general staff to be dissolved ;

- Pending agreement on war reparations, Germany was to make a provisional payment of 20,000 million marks in gold;
- German foreign financial holdings to be confiscated;
- Under Article 231, Germany and its allies were specifically blamed for the war and held responsible for all damage and losses suffered by the Allies.

Not surprisingly, these terms were greeted with universal protests within the Reichstag. In the end, again not surprisingly, the verdict was general and unconditional acceptance.

One week later, on the 28th, the Allies and their humbled foes signed the Treaty of Versailles. Unfortunately for sticklers of international law, the U.S. Senate declined to ratify the Treaty, or indeed a Treaty of Alliance between the U.S., France and Britain providing for military assistance in the event of a further attack from Germany.

2

Postwar Disasters

The delayed consequences of the Treaty of Versailles were the ascent to power of Hitler and his National Socialist (Nazi) party; and the outbreak of World War II, twenty-one years after the end of World War I. Another way of putting it would be to point out that, in sum, the conditions imposed by the victorious Allies proved to be counter-productive.

Which leads to a more specific point. In a major war, such as the Great War—the biggest and most destructive war in history at that time—the most important thing to grasp is that in a dictatorship or an authoritarian regime, the people are not necessarily responsible for the horrors inflicted by the armed forces. It is undoubtedly necessary to destroy the military machine of the losing power, when it is known that the said power started the war. It may also be desirable to punish not only the military, but also the unelected politicians whose ambitions sparked the conflict. However, it is (in my view) both morally wrong and ethically undesirable to inflict excessive suffering, notably through deliberate deprivation, on a powerless population.

After the defeat, chaos reigned for a few weeks. Hoping to placate the victorious allies, Ludendorff ordered that Germany should become a constitutional monarchy. As mentioned in the last chapter, General Ludendorff had not, at this stage, read President Wilson's Fourteen Points. Now, with defeat closing in, he caught up on his required reading, and was shocked. So shocked, indeed, that he let it be known that he wanted to resume fighting. In this, he appeared isolated, so he resigned rather than recognize that peace had come.

On 4 October 1918, Prince Max of Baden, well known internationally for his liberal views, was named as chancellor. That same day the political leaders were told what they knew already, that Germany had lost the war.

13

On 28 October, sailors at Kiel were ordered by the Admiralty to return to sea and fight the British. Instead, they started a mutiny, and shouted a defiant phrase: "Further than Heligoland we will not go." Their stand of defiance proved contagious. On the 29th, Emperor Wilhelm II, alarmed by Reichstag demands for his abdication, left Berlin for army headquarters at Spa. By 4 November, the Kiel revolt had spread to other seaports, and revolutionary councils of workers and soldiers were formed. There followed, in fact, more than one revolt. Three days later an independent socialist named Kurt Eisner led a revolt in Munich, followed on the 8th by the proclamation of a (short-lived) republic in Bavaria.

On the 8th the cabinet of Prince Max of Baden advised the emperor to abdicate, to save the nation from civil war. The Emperor, still clinging to a non-existent power, rejected the advice. His defiance did not last long, however, for the following day (9 November) he did what he had said he wouldn't do: he abdicated and fled to the Netherlands. H is decision was announced in Berlin.

That same day, Prince Max had ceded the chancellorship to the Socialist leader, Friedrich Ebert, And again, that eventful day, a Social Democrat, Philipp Scheidemann, proclaimed a republic, in order to forestall the proclamation of a Soviet Republic by Karl Liebknecht, leader of an extreme left group who styled themselves the "Spartacists."

On the 10th, Ebert became chairman of the Council of People's Representatives. His Social Democratic Party, which opposed revolution, dominated the new Chamber. Anxious for a legal and democratic solution to Germany's postwar problems, he reached a bargain with Hindenburg, under which the latter would retain his command while Ebert would oppose revolution. The two men agreed that elections for a National Assembly would take place on 19 January 1919.

A fortnight or so before this date, however, street fighting broke out in Berlin, as Liebknecht and his political partner Rosa Luxemburg tried but failed to stage a revolution. Both were arrested, then murdered (rather than executed) on 15 January 1919.

When the new National Assembly met at Weimar on 6 February that year. Ebert was elected president of the Reich, with the Social Democratic leader, Philipp Scheidemann, as Chancellor.[1] But anarchy still reigned. On the 21st, Kurt Eisner was assassinated by plotters intent on restoring the monarchy. In February and March there were further Communist uprisings in Berlin and Munich. Acting for the new government, Gustav Noske suppressed these revolts. But anarchy continued: on 4 April, a "Soviet Republic" was proclaimed in Bavaria. By 1 May it was overthrown by the armed forces of the federal government. Nor was that all: on 1 June, another Republic was proclaimed, this time by a self-styled Rhineland Republic, apparently plotted and supported by Germany's victorious enemies, the French. Not unexpectedly, the inhabitants of the new "Republic" were hostile and, after a few months, the alleged Republic collapsed.

On 20 June, the Scheidemann ministry resigned, as a protest against signing the Peace Treaty dictated by the victorious Allies. One unsatisfied protester was Admiral Ludwig von Reuter, commander-in-chief of the German navy. His protest was dramatic and irreversible. On 21 June, he ordered the total scuttling of the German warships by their crews. The ships that sank included ten battleships, nine armored cruisers, eight smaller cruisers and 102 submarines, for a total of about 500,000 tons. This dramatic act of protest had been made possible by the fact that the Allies had insisted on the internment of the whole fleet at Scapa Flow.

The powers conferred by the election were, in some respects, greater than those wielded by the defunct monarchy. All taxation was under the control of the central government, whose authority overrode that of the individual states (*Länder*). Each Land was represented in the Reichsrat, but this chamber was subordinate to the Reichstag, chosen by all men and women over the age of twenty, on a system of proportional representation.

Supreme power, however, was vested in the president. Elected for seven years by national vote, he was eligible for reelection. He was also the supreme commander of the armed forces and his powers included that of dissolving the Reichstag, and submitting any law with which he disagreed to a referendum. Moreover, under Article 48 of the new Constitution, he was entitled to suspend civil liberties in an emergency.

There were now seventeen *Länder,* or states. The smallest was Schaumburg-Lippse with a population of merely 48,000 to Prussia, with 38 million inhabitants (as measured in 1925). One of the *Länder* was new: Thuringia, an entity formed by amalgamating seven small principalities. All of them were represented in the lower Chamber—the Reichsrat—which in fact was subordinate to the Reichstag, to which the Chancellor and his government were responsible. All men and women over twenty were entitled to vote representatives into the higher Chamber. However, as a kind of counterweight to the Reichstag, the president, with full powers as chief executive, was to be voted into power by direct national vote for a seven-year term, and would be eligible for reelection. His authority extended to making alliances and signing treaties. Moreover, he would be the supreme commander of the armed forces and his authority would extend to dissolving the Reichstag and submitting any law passed by it to a referendum. Nor was this all: he also had the right, under Article 48, to dissolve the Reichstag and suspend civil rights in the event of an emergency.

In its favor, the new German Republic was democratic but not Socialist.

Despite the vast powers conferred by the Constitution, the successive Weimar governments were frustrated by both internal challenges and foreign pressures. For the first time, Germany and the outside world hear or read the name of Adolf Hitler, and his National Socialist German Workers' Party in Munich. Ex-army officers were providing arms to former servicemen joining the Nazis.

Outside pressures were mounting. On 27 April 1921 the Allied Reparations Commission presented the new Republic with a bill for132 billion gold marks. The demand broke the central government. Meanwhile, pressures from the neighboring French government were mounting. When Germany defaulted on timber deliveries, the French forces retaliated by occupying the Ruhr, in January 1923.

An unpredicted move followed. Harassed by the Western Allies, and the French in particular, the Weimar government successfully negotiated a deal with Lenin's fledgling government in Russia. On 16 April 1922, the two outcast governments signed the Treaty of Rapallo. It was a treaty of "friendship," providing for an exchange of Russian raw materials for German technology. The full significance of the deal emerged over the next few years, culminating in the Treaty of Berlin on 24 April 1926, enshrining friendship and neutrality between the two countries. Secret arrangements followed between the German and Soviet High Commands, which gave the Germans facilities for training on Soviet soil in the use of armaments secretly manufactured in Germany. Naturally, the new Soviet Red Army benefited equally from these joint facilities. The German use of them was of course in clear violation of the Treaty of Versailles.

In the decade of 1923 to 1933, the biggest problem for defeated Germany was inflation. The main cause of it was inevitably the commitment of successive governments to meet the Versailles Treaty's provision for reparations. In November 1923, the then chancellor, Gustav Stresemann, launched a new Rentenmark, to be covered by mortgages on the country's entire agricultural and industrial production.

Sensible though this arrangement appeared, it angered the extreme Right, as well as the Communists, and for the first time brought international headlines to the Nazi leader, Hitler. The Communists attempted a rising in Hamburg. Hitler's badly organized coup became known as the "beer hall Putsch," the aim of which was to overthrow the Bavarian government, Surprisingly, perhaps, Field Marshal Erich Ludendorff was also involved. The occasion the conspirators had chosen was a lecture in the Bürgerbraukeller, the biggest beer hall in Munich, on the moral justification of a political dictatorship.

The lecturer had hardly opened his mouth when a gang of men wearing brown shirts burst in. They were led by a man with a toothbrush moustache who jumped on a chair, fired a shot into the ceiling and shouted: "The national revolution has begun." He claimed Ludendorff's backing. Fetched from his home outside the city, Ludendorff berated Hitler for starting the coup without him. He then tried to lead a march but was stopped by police and put under house arrest. Hitler fled but was also arrested. The trial of the conspirators, however, did not take place until 1 April 1924, an appropriate date, in that Hitler was allowed to interrupt witnesses and make speeches about Karl Marx and the great (but anti-Semitic) composer, Richard Wagner. He was sentenced to five years in jail, but released after less than a year. While in prison, Hitler wrote his sensational book *Mein Kampf* ("My Struggle"). His fellow conspirator, Ludendorff, was acquitted.

For a while, the Rentenmark remedy appeared to work, and in August 1924 the national Reichsbank was made independent of government control and a new currency, the Reichsmark, was launched. One new Reichsmark was worth 1 billion (1,000,000,000,000) of the marks then in circulation. The immediate effects were dramatic: unemployment fell as production boomed.

The problem of reparations remained unsolved, however. A committee under the chairmanship of an American financier, Charles G.

Dawes, produced a report on 9 April 1924, under the slogan "Business, not politics." The Dawes Plan, as it became known, provided for the reorganization of the Reichsbank under Allied supervision. The main provision was for reparation payments of one (American) billion (1,000,000,000) gold marks a year, rising to 2,500 million after five years. As if to underline the absurdity of the arrangement, it was agreed that Germany should receive a foreign loan of 850 million Reichsmarks. Five years on, another committee, on which Germany was represented, chaired by the American Owen D. Young, fixed the total of German reparations at 121,000 million Reichsmarks, to be paid in fifty-nine annual instalments.

The net effect of these monetary measures was a virtually uncontrollable inflation. On 26 April 1924, Hindenburg was elected president of the Republic, with a narrow majority over the centrist prime minister, Wilhelm Marx (14,655,766 over 13,751,615). In a speech three years later (18 September 1927), President Hindenburg repudiated German responsibility for the war, until then enshrined in Article 231 of the Versailles Treaty.

Two years on, an almost apocalyptic event thousands of miles away caused a panic wave in many countries, not least in debt-burdened Germany. "Black Thursday," as it became known, was the label given to the great Wall Street crash on 24 October 1929. Confusion and panic reigned as 13 million shares changed hands on the New York Stock Exchange. As reports put it, "dazed brokers waded through a sea of paper clutching frightened investors' orders to 'sell at any price'." Had Karl Marx still been around, he would have gloated over what he would have seen as the predicted collapse of capitalism.

The crash sparked off a worldwide depression, which affected many countries, but none as severely as Germany. Banks started tumbling, not least the sizeable Darmstätter und Nationalbank in July 1931. Early in 1932, unemployment passed the 6 million mark.

These developments suited Hitler, who had made it clear that his ambition was to win the chancellorship by legal means, not by revolution.

The going was tougher than he had imagined, however. On 13 April, the government imposed a ban on the Nazi storm troops, but on 16 June the ban was lifted. Meanwhile, on 4 June, the Reichstag was dissolved. On 20 July, martial law was proclaimed in Berlin and Brandenburg after Nazi demonstrations and marches had in effect ruled out civil order. Further elections were held on 31 July, when the Nazis won 230 seats, to 133 for the Socialists and 89 for the Communists. The Centrists won a mere 97.

Both the Nazis and the Communists refused to enter a coalition; which ruled out a governing majority. When on 13 August, a desperate President Hindenburg asked Hitler to serve as vice chancellor under the Centrist Franz von Papen, Hitler refused. He stated his unconditional aim: "All or nothing"

The Reichstag was dissolved on 12 September, and Hitler refused another conditional offer, this time of the chancellorship. His hour came on 30 January 1933, when President Hindenburg offered him the chancellorship, unconditionally as regards policies, though with von Papen as vice chancellor. The extreme Right revolution followed. And so,too, did World War II, on 1 September 1939, when Hitler's army invaded Poland.

3

The Third War (1939-1945)

Democracy had brought Hitler to high office, and democracy would give him the full powers he needed to take Germany's revenge on the Western world. His first taste of power, however, must have been frustrating, for his Nazis had been allotted only three of the eleven ministries. Moreover, apart from his own chancellorship, these were not key posts. One of them, however, turned out to be a key Nazi weapon: the "job description" was minister of the interior of Prussia. It went to Goering and gave him control of the Prussian police); Goebbels (Hitler's master of propaganda) was left out. At that time, popular support for the Nazis was as low as 37 percent, but the 63 percent who opposed them were deeply divided. The main rival force was the Communist Party which, however, had a visibly silly plan: to eliminate the rival Social Democrats first.[1] The first move Hitler took, on Goering's suggestion when his cabinet met for the first time, was to persuade his ministers to support his call for new elections. His one object was to achieve a majority in the Reichstag, so that whatever policies he wanted would be supported by a majority vote.

He got his way and the date set for the second elections he needed was 5 March 1933. Goebbels was jubilant. as his diary entry for 3 February would confirm: "Now it will be easy to carry on the fight, for we can call on all the resourses of the State. Radio and press are at our disposal. We shall stage a masterpiece of propaganda.[2]

And this was precisely what happened. Goering set up a Nazi-controlled police force of 50,000 men. He staged a raid on Communist Party headquarters and claimed, on the basis of seized documents, that the Communists were planning to launch a revolution.

A sensational news item on 27 February played into the hands of the Nazis: the Reichstag was on fire. Who lit the fire may never be known, but what is certain is that the flames served Nazi purposes. The government immediately assumed emergency powers, but despite this the Nazis failed to gain an overall majority when the elections took place a week later. The Nazis won 288 seats in the new Reichstag, but persuaded fifty-two Nationalist members to support them, which gave Hitler a bare majority of the 647 newly elected deputies.

Hitler's next move was the Enabling Act, which was passed when the Social Democrats were outvoted by 441 to 49 after the eighty=one Communist deputies had been arrested or in other ways neutralized. The Enabling Act soon emerged as Hitler's "constitutional" key to total power. The following decisions sum up the Nazi revolution:

- As minister of the interior of Prussia, Goering fired hundreds of officials and replaced them with Nazis, most of them Sturmabteilung (SA: storm troopers, better known as Brownshirts) or Schutzstaffel (SS: Blackshirts).
- The Land diets were abolished by a series of decrees, transferring their powers to the Nazi government.
- Trade unions were abolished in May 1933, by a merger into a single Labor Front.
- All political parties except the National Socialists were suppressed.
- Goebbels was brought into the cabinet as minister of public enlightenment and propaganda.
- Von Papen was ousted as commissioner for Prussia and replaced by Goering.

By this time, in the spring of 1934, tensions were mounting between the German army and the Nazis. The inevitable crisis came in June when Hitler visited the dying president, Marshal von Hindenburg, in his eighty-seventh year. Also present was General Werner von Blomberg, minister of defense. The general was in an uncompromising mood. He warned Hitler that unless the government managed to relax prevailing tensions straightaway, Hindenburg would hand over full powers to the Army.

Hitler immediately discussed the problem with Goering and Himmler. The three of them agreed that the best way to bring the crisis under control was to get rid of Ernst Röhm, head of the SA. Some years earlier, in 1925, Hitler and Röhm had parted company after a row, and Röhm had emigrated to Latin America and found a job in the Bolivian Army. Now, in 1934, he headed the SA. Hitler gave the job of ridding the party of Röhm to Goering and Himmler, who headed the SS. Their solution was suitably extreme. Himmler in particular told the Führer that Röhm was plotting a *putsch* against him. Hitler ordered Himmler to do the necessary. On 30 June 1934, Röhm and several of his lieutenants were seized on Himmler's orders and executed without trial. Altogether, seventy-seven persons were executed in what became known as the Great Blood Purge.

Hindenburg died on 2 August and the Army, delighted by the removal of Ernst Röhm, raised no objections when Hitler assumed the dual title of Führer und Reichskanzler ("Leader and State Chancellor"), although he opted to use only the "Leader" one. He was confirmed in his new office by a plebiscite on 19 August.

From that point on, Hitler's team completed the transformation of Germany into a totalist state. All existing organizations were either abolished or transformed into instruments of Nazi rule. The schools and universities were indoctrinated and the pervasive Hitler Youth movement made sure the young of both sexes were turned into good Nazis. The Nazi regime harassed both Catholics and Protestants. In April 1933, Jews had been dismissed from all official posts and from universities, and debarred from entering all professions. In November 1938, on Hitler's orders (via Himmler), a pogrom was carried out; most Jewish property was confiscated and surviving Jews were confined to ghettoes.

While Nazifying Germany, Hitler made sure that the economy was in safe hands, by appointing the head of the Reichsbank, Hjalmar Schacht, as his chief economic and financial adviser. Schacht played a leading role in the financing of German rearmament. However, on 2 January 1939, his financial orthodoxy incurred Hitler's displeasure, when he urged the Führer to reduce expenditure on armaments as the only way to prevent inflation and balance the budget. By that time, Hitler's preparations for the forthcoming war were advanced and Hitler was in no mood to listen to "sensible" advice of this kind. So Schacht was sacked as economic adviser (although he stayed on as head of the Reichsbank).

While the Nazi revolution was in progress, Hitler gave repeated signals to the outside world that he intended to spread German power over the whole of Europe. The most relevant events were these:

- On 26 January 1934, Germany and Poland concluded a Treaty, providing for a ten-year period of non-aggression and respect for existing territorial rights. This brought to an end a period of extreme tension.

- On 25 July of that year, Chancellor Dollfuss of Austria was assassinated during a Nazi coup. A period of German-Austrian tension followed.
- Under the Treaty of Versailles, the League of Nations conducted a plebiscite in the Saar Basin on 13 January 1935. Ninety percent of the voters opted for reunification with Germany. This was the first step in German expansion under the Hitler regime.
- On 16 March of the same year, Hitler denounced the clauses of the Versailles Treaty calling for German disarmament. Surprisingly, Fascist Italy (which Hitler had visited the previous year) joined France and Britain in a strong protest against Hitler's move.
- On 18 June that year, Germany reassured the British by promising not to expand its navy beyond 35 percent of the British strength. The naval agreement not only reassured the British but drove a wedge into the prevailing Franco-British entente.
- On 7 March 1936, Hitler's government denounced the Locarno Pacts of 1925 and reoccupied the Rhineland.
- On 11 July, a German-Austrian agreement ended a period of high tension between the two countries.
- Mussolini's foreign minister, Count Ciano, visited Berlin
- between 25 and 27 October 1936. The outcome was the Berlin-Rome Axis, seen as a challenge of the "have-nots" to the "haves": France and Britain.
- 14 November 1936. Hitler denounced the clauses of the Versailles Treaty calling for international control of German rivers.
- 25 November. A German-Japanese pact was concluded, Ostensibly directed against communism, it was seen in reality as an extension of the Berlin-Rome Axis.
- 12-13 March 1938. Germany invaded and annexed Austria, thereby increasing Germany's population by 6 million.
- 12-29 September. Hitler annexed the German-populated Sudeten region of Czechoslovakia, bringing 3 million more German speakers under his control. On 15 March 1939, Hitler added rump Bohemia and Moravia to the Czech territories already seized.

To the astonishment of the Western powers, Hitler concluded a Pact with Stalin on 20-21 August 1939. On 1 September, without a formal declaration, the German forces crossed the Polish border. World War II had begun. Two days later, Britain and France declared war on Nazi Germany.

A retrospective summary would be useful. The regime that emerged from Germany's defeat in the First World War is perhaps best described as an inexperienced democracy. The political parties that emerged, although not responsible for the hostilities, were held responsible for the losses suffered b the Western powers. The crushing terms imposed on the nascent democracy built up a deep and demoralizing atmosphere, while the economic consequences were too damaging to the German people to bring the expected benefits to the other Western populations.

Moreover, in the absence of a postwar allied occupation of German territories (except the Ruhr and the Saar), the new German administration was, in effect, held responsible for the errors and crimes of the defeated one. As mentioned earlier, the Wall Street crash of 1929 hit Germany considerably harder than the Western powers.

All these circumstances played into the hands of the vengeful founder and leader of National Socialism, Adolf Hitler. The unplanned consequences of German democracy included the advent to power of a fanatical enemy and exploiter of that democracy.

Fortunately for the Western world in particular, and for the world beyond the West in general, the victorious allies did not repeat the errors and blunders of the post-1918 peace. To be more precise, this observation applies to the Western allies (the U.S., Britain, and France) and not to their fortuitous ally, the Soviet Union and its absolutist dictator, Josef Stalin.

In the context of this book, by far the most important event of the immediate postwar period was the Potsdam conference, from 17 July to 2 August 1945. The chosen place was a hotel, the Cecilienhof, in Potsdam, in effect, a suburb of Berlin, largely destroyed by Allied bombing. The participants were the political leaders of the three major wartime allies, and their foreign ministers: Generalissimo Josef Stalin and Vyacheslav Molotov for the USSR; President Harry Truman and Secretary of State James Byrnes for the United States; Prime Minister Winston Churchill and Foreign Secretary Anthony Eden for the United Kingdom. Discussions were suspended on 25 July because of the first postwar elections in Britain. When they were resumed, on the 28th, the British participants were Clement Attlee (who had replaced Churchill as prime minister), and Ernest Bevin, the new foreign secretary.

The most important decisions of the Potsdam conference were those affecting Germany:

- Disarmament and demilitarization;
- Dissolution of Nazi institutions;
- War criminals to be brought to trial;
- Encouragement of democratic procedures;
- Freedom of speech, press and religion (subject to the requirements of military security).

Other decisions concerned measures for Germany to compensate for losses and sufferings caused to other nations, and the following economic restrictions:

- Prohibition of the manufacture of war materials and weapons;
- Controlled production of metals, chemicals and machinery essential to war;
- Decentralization of German cartels, syndicates and trusts; encouragement of agriculture and peaceful domestic industries; and
- Control of exports, imports and scientific research.

The most important decision of all, however, was the division of Germany into four occupation zones. By far the largest was the area overrun by the Soviet forces in the last phase of the war, which included a large slice of Polish territory. Moreover the German capital, Berlin, although also divided into occupation zones, happened to be surrounded by Soviet territory. The second in area was the American zone and the British came third. A more modest French zone was carved out of the British and American zones.

An acceptable form of representative democracy was introduced with remarkable speed in the Western Allied zones. With Allied encouragement, the emergence of democratic parties was encouraged and elections took place in the U.S. zone in June 1946, in the British zone in April 1947, and in the French zone in May 1947. The main parties that emerged were the Christian Democratic Union (CDU), the Free Democrats (liberals) and the Social Democrats (who included the trade unions).

Initially, in the postwar period, defeated Germany lacked a government. Instead, an Allied Control Council, consisting of representatives of the Soviet Union and the three Western allies, gave the orders. On 20 November 1945, the trial of the remaining major Nazi leaders (except Hitler, Himmler, and Goebbels, who had opted for suicide) opened at Nuremberg. Most of the twenty-two defendants were sentenced either to long terms of imprisonment, or to death. Of the latter, the most prominent was Goering, who committed suicide on 15 October, shortly before his scheduled execution.

Not surprisingly, it was immediately clear that whatever the degree of harmony between the Western allies, Stalin's Russia would take its own retributive measures. The United States and Britain had learned the lessons of the post-World War I period. It would be to everybody's advantage to guide West Germany towards a self-supporting economy. Stalin opted for punishment, namely $10 billion in German reparations. This fundamental divergence emerged during the Moscow conference of the Big Four foreign ministers in Moscow from 10 March to 24 April 1947. Already, on 2 December, James Byrnes and Ernest Bevin had signed an agreement for the economic merger of the U.S. and British occupation zones, to be known as "Bizonia." France and the Soviet Union were invited to join; France accepted, Russia did not.

By then, it had become clear that the wartime alliance of convenience between the Communist superpower and the democratic West had ended. On 1 April 1948, the Soviet occupation force began to interfere with traffic between Berlin and West Germany. The climax came on 24 June, when the Soviet authorities ordered a complete stoppage of Western traffic to and from Berlin, whether by road or by rail. To their credit, the Western allies opted for travel by the remaining option: flying. The Soviet blockade endured until 30 September 1949, after no fewer than 277,264 allied flights The counter-blockade had worked.

Meanwhile, epoch-making events had taken place. On 23 May 1949, the Western powers authorized the launching of the Federal Republic of Germany, On 14 August elections were held in West Germany for the Bundestag or lower house, producing a small majority for the Christian Democrats over the Socialists, with the balance held by the Free Democrats. In mid-September, a Free Democrat, Theodor Heuss, was elected president, and the Christian Democrat leader, Konrad Adenauer, as chancellor (prime minister) of the new Federal Republic.

On the Soviet-controlled side, parallel but deeply divergent events had been taking place. On 16 May 1948 elections for a "People's Congress had taken place in East Germany. The resulting majority for the single list of Communist-approved candidates was of course inevitable, but the protesters boycotted the polling in considerable numbers, so that the Communist majority was only 61.1 percent.

There is no need to recount the history of divided Germany in all its details. The essential points (in the context of this work) are these:

- Under the wise guidance of the Western powers, West Germany grew into one of the major success stories of the twentieth century, economically, politically, and strategically. An indirect consequence of Communist North Korea's invasion of South Korea in 1950 was the transformation of West Germany from foe to friend, or more precisely from enemy to ally.

- On 19 September 1950, the Western powers announced that any act of aggression against the German Federal Republic or against West Berlin would be treated as an attack upon themselves.
- On 26 May 1952, the internal independence of West Germany was recognized by the United States, Britain, and France.
- On 5 May 1955, the German Federal Republic achieved sovereign status.
- Meanwhile, the Soviet Union consolidated its grip on the East German "Democratic Republic."The rivalry between the two Germanies and their divided capital culminated in Moscow's decision, in August 1961, to build a wall to divide East and West Germany. Twenty-eight years on, on 10 November 1989, the Wall was breached from the eastern side and a crowd estimated at 2 million poured into West Berlin. In Bonn, the West German Chancellor, Helmut Kohl, seized the moment and brought the two Germanies together. On 31 August 1990, a 900-page treaty of reunification was signed in East Berlin. The Cold War had ended and communist East Germany, impoverished by the long years of Soviet domination joined prosperous West Germany.

The Two Leaders: Adenauer and Erhard

Konrad Adenauer was a great example of a relatively rare political: phenomenon: the late developer. Had he died in 1945, aged sixty-nine, his career would have merited no more than ten or twelve lines in international encyclopaedias. Instead, he became a major leader of his own country–Germany–and one of the leading architects of the Western Alliance to hold the Soviet threat at bay. He died in 1963, aged eighty-seven, his place in the history books well merited and permanent.

Born in Cologne and brought up as a Roman Catholic, he was elected to the Cologne municipal council in 1906, and went on to be Oberbürgermeister (in effect Lord Mayor) in 1917. He remained in local politics through the Weimar Republic. Under Nazi rule, he was jailed twice, for alleged political offences. He felt his hour had come when he returned to Cologne after the war, with one dominant thought in mind: to rebuild and refurbish his bomb-shattered birthplace. No such luck: this time, he was removed by Britain's military governor of the North Rhine area, Sir John Barraclough. His offence? "Politically incompetent"!

He found his salvation and his future when the Christian Democratic Union was formed, bringing Catholics and Protestants together.

His hour came when he was elected chairman of the CDU in August 1949. The CDU narrowly defeated the Social Democrats (SDP) in West Germany's first general elections in August 1949, and Adenauer became chancellor. He held his post until, aged eighty-eight, he handed it over to Ludwig Erhard, who had been his Eonomics Minister from 1949 to 1963.

If Adenauer had guided the new Germany to its place in the postwar world, it was Erhard who had rebuilt his country's economy from its inflationary postwar misery, transforming it into one of the most prosperous countries in the world. The two men did not get along, In Adenauer's eyes, Erhard, despite his success, was politically too "free and easy." Nevertheless, Adenauer did not quarrel with the view, generally accepted, that Erhard was his natural successor.

Not only did Adenauer preside over postwar Germany's advent to democracy, he also cooperated with the French leader General de Gaulle's covert action plan to foster a deep-seated partnership with France's enemy of three wars, which became the real and lasting heart of the European idea.

On the French side, de Gaulle's choice to guide this major initiative was the popular politician Antoine Pinay, who was France's economics minister in 1949, prime minister and finance minister for most of 1952 and foreign minister in 1955-56; and de Gaulle's finance minister in 1958. As de Gaulle himself put it, he had chosen Pinay to restore economic and financial stability in France because "this eminent personage, well known for his common sense, esteemed for his character, popular for his devotion to the public interest," would strengthen confidence and help avert the threatening catastrophe.[8] It was a sound choice. On Pinay's advice, de Gaulle launched a national loan that was spectacularly successful.

This was the public Pinay. But there was also a secret Pinay, the man chosen by de Gaulle to visit Chancellor Adenauer and engineer a firm but discreet relationship that would form the heart of the European movement and forever remove the risk of any further Franco-German wars.

Specifically, Pinay's secret mission was to brief the German chancellor on the French leader's policies and to propose the creation of a secret Franco-German friendship organization, which became known as Le Cercle Pinay. For years, prominent French and German leaders in various walks of life met in secret twice a year. The initiative was in the hands of a Paris lawyer, Jean Violet, who invited me to join the Cercle in 1970. Already, in 1958, I had been invited to join a parallel organization, similarly motivated. Both were controlled by the French foreign secret service, known as the SDECE (Service de Documentation extérieure et de Contre-espionnage, the French equivalent of Britain's MI-6). The other SDECE organization, run from a modest private office in Paris by Antoine Bonnemaison through an organization calling itself the Centre de Recherche du Bien Politique ("Research Centre for the Political Good") complemented the Cercle.

It, too, ran twice-yearly meetings called Colloques, initially limited to French and Germans but later broadened to include other Western countries.[9]

The point to note is that French officialdom, before and after the return of de Gaulle to power in 1958, devoted a good deal of time and money to developing and maintaining close relations with their German neighbors. The era of Franco-German military conflicts had been brought to an amicable close: a perfect example of political victory.

4

MacArthur's Japanese Legacy

General Douglas MacArthur was more than a great military leader: he was also a great statesman who did more than anyone, or any government, to transform militarist Japan into a prosperous and peaceful power. The Western world, as well as Japan, owes him a debt of gratitude.

In a special sense, the story of Japan's rise to great power status began in the middle of the nineteenth century with the arrival of foreigners (dismissed as "barbarians") on Japanese soil. At that time, Japan was still a feudal country, after 700 years of feudalism. The turning point came in February 1868. with the advent to power of a new and vigorous emperor, Mutsuhito, who assumed the name of "Meiji," meaning "enlightened rule." Feudal lands were returned to the throne, and in effect Japan embarked on a new age of modernization, marked by rapid industrial and economic expansion, and a dramatic growth of population: from 30 million at the time of the Meiji Restoration to nearly 65 million in 1930.

With modernization came wars, notably with China (1894-1895) and with Russia (1904-1905). In April 1895, China recognized the independence of Korea and ceded the island of Formosa (Taiwan) to Japan. Korea was formally annexed to Japan in August 1910.

Between 1931 and 1939, Japanese military expeditions brought a considerable area of China under Japan's control. In 1940 the Japanese also gained control over much of the French colony of Indochina. It was during this period that a major player in the Japanese army rose to prominence. His name was Tojo Hideki. By 1937, he had become chief of staff of the Japanese Kwangtung army in Manchuria.

On 27 September 1940 Japan joined Germany and Italy in a ten-year Tripartite Pact "for the creation of conditions that would promote the prosperity of their peoples." By that time, Tojo had been appointed war minister by the then prime minister, Prince Konoe. In October 1941, Tojo successfully forced the resignation of Konoe, whom he replaced as prime minister, while remaining war minister.

A major point at issue between Tojo and Konoe had been Japan's relations with the United States. The U.S. had frozen Japanese assets and placed an embargo on American exports of oil to Japan. The Japanese government was faced with an unattractive choice: withdrawal from Indochina, and possibly China, or holding on to its conquests while seizing the Netherlands oil wells and production in the Dutch East Indies. The U.S. rebuffed a request from Konoe for a meeting with President Roosevelt unless the Japanese first met American requests for concessions. It was at this point that Tojo forced Konoe out of office and stated Japan's "final offer" to the U.S. Secretary of State, Cordell Hull: Japan would pull out of Indochina, but not until America agreed to resume oil shipments, cease aid to China and unfroze Japanese assets.

By then Tojo was preparing for a military attack on America, but instructed his diplomats to continue "peace talks" with the Americans. The attack came on 7 December 1941, when Japan's sea and air forces launched devastating surprise attacks on the U.S. base at Pearl Harbor, Hawaii, and on British forces in Hong Kong and Malaya. The U.S. forces were caught totally unprepared. Five battleships and three cruisers were sunk or seriously damaged, three battleships less seriously damaged, many smaller vessels sunk or crippled and 177 aircraft destroyed. The American casualties were heavy: 2,343 killed, 876 missing and 1,272 injured.

Next day the U.S. declared war on Japan, but the Japanese stuck to their carefully organized battle plan. Their air and naval forces attacked Guam and Wake Island. On 13 December, Guam surrendered, and seven days later so did Wake Island. On the 20th, the day of Wake Island's surrender, Great Britain declared war on Japan; justifiably, for on 10 December, the British battleship *Prince of Wales* and the battle cruiser *Repulse* had been sunk in a Japanese air attack.

At this early stage of the Pacific War, the odds seemed to favor Japan. On 25 December 1941 they had taken Hong Kong; on 2 January 1942 they captured Manila; and on the 11th they started occupying the Netherlands East Indies. On 15 February they seized Singapore from the north, whereas the British had supposed, wrongly, that the attack would come from the sea to the south. On 7 March they occupied Rangoon and spread their forces over Burma.

The run of successes would not last, however. On 7 May of that same year, Allied naval and air power destroyed 100,000 tons of Japanese shipping in the Battle of the Coral Sea, thus frustrating a Japanese plan to occupy Australia.

The U.S. naval buildup was gathering strength. On 7 August U.S. Marines landed in the Solomon Islands, capturing Japanese airfields on Guadalcanal. On 12 November the U.S. Navy crowned a three-day battle in the Solomon Islands, sinking one Japanese battleship, five cruisers, and twelve transport ships.

The years 1943 and 1944 were marked by one Allied victory after another. A selective list of Japanese defeats follows:

- 22 November 1943. U.S. forces land on the Gilbert Islands, crushing Japanese resistance in three days.
- 2 February 1944. U.S. forces invade the Marshall Islands.
- 16 June. U.S. Superfortress planes raid the Japanese home island of Kyushu.
- By July, U.S. bombers were within range of Tokyo. There was an important political consequence to the turn of events: Tojo was forced to resign, to be replaced by a milder man, Koiso Kuniaki.
- 11 August. U.S. forces reconquer Guam. The last Japanese invaders are driven back to Burma across the Indian frontier.
- 19 October. General MacArthur commands U.S. invasion forces on the island of Leyter, to start the invasion (or liberation) of the Philippines.
- 25 October. What was left of the Japanese fleet withdrew from Philippine waters: forty ships had been sunk and forty-six damaged; 405 planes had been smashed. Japan faced defeat but still refused to surrender. The great battle for Iwo Jima, one of the Pacific Volcano Islands, seen by MacArthur as the key to the invasion of Japan, was seized by the Americans on 19 February 1945, but at a high cost: 4,000 dead and 15,000 wounded.
- 1 April. American marines and soldiers invaded the southern Japanese island of Okinawa. On 7 April, U.S. aircraft sank the Japanese battleship *Yamato*, plus two cruisers and three destroyers. Meanwhile, in Southeast Asia, the 14th British Imperial Army, commanded by Admiral Lord Louis Mountbatten, with support from Chinese and U.S. forces, completed the destruction of the Japanese 15th, 28th and 33rd armies. The fighting had lasted 15 months, and the Allied victory was announced on 30 April.

Back to Okinawa. On 5 May, the Japanese the Allies released their count of Japanese casualties, at 347,000. The Japanese defenders went on resisting until 21 June, but by then the island had become a major airbase for direct raids on Japanese cities.

From May to August, the United States carried out the greatest air offensive ever, destroying or immobilizing what was left of the Japanese Navy, shattering Japan's industry and shelling Japan's most densely populated cities with impunity.

But there was worse to come. Japan's civilian population was about to be the target of a revolutionary scientific advance in destructive technology, the product of years of secretive American and British research: the atom bomb. On 6 August, an American Superfortress B-29, flying alone at 30,000 feet, dropped a single atom bomb on the city of Hiroshima. Four square miles of houses were destroyed and some 80,000 inhabitants were killed. If the number of patients affected by radioactivity who died within a year are added to the immediate casualties, the total killed by this single attack reached about 140,000.

Three days later, on 9 August, the U.S. Air Force dropped another atomic bomb, this time on the seaport of Nagasaki. Casualties were lower, but still searingly massive: 24,000 immediate deaths, followed by 30,000 hospital patients, most of whom would die.

This was it: Japanese resistance had ended and on 10 August the Japanese offered to surrender. The Allied terms had been prepared and were communicated to Tokyo. They were accepted four days later, on 14 August.

On the 26th, U.S. occupation forces landed in Japan and the formal terms of surrender were signed on board the *USS Missouri* in Tokyo Bay on 2 September 1945.

A dramatic but unexpected event took place on 8 August, between the first and the second atomic bomb attacks on the two Japanese cities: on Stalin's orders, the Soviet Union declared war on Japan and invaded Manchuria. "Unexpected" applies to the broad Western public at the end of World War II. Some months earlier, at the Yalta conference between the Soviet leader, and the U.S. and British leaders, President Roosevelt and Prime Minister Churchill, Roosevelt in particular had pressed Stalin to enter the war against Japan. Stalin had agreed, but only if certain conditions were met. These included the restoration of territories seized by Japan in the Russo-Japanese war of 1904-1905. Roosevelt had accepted Stalin's conditions; and so, a day or two later, had Churchill.

By 14 August, only six days after the Soviet invasion, Manchuria was under Soviet control, and the Soviet forces, licensed to steal, plundered Manchurian territory of food, gold, and industrial machinery to a tune of uncounted billions of dollars.

In his victory speech to the Soviet people on 2 September 1945—the day of Japan's formal surrender on board the *Missouri*—Stalin justified the invasion in a speech referring to Russia's 1905 defeat by Japan, which had "left a bitter memory on our minds, for it stained our name." [1]

After Japan's defeat General MacArthur was named Supreme Commander for the Allied Powers (SCAP). In effect, however, he emerged as a benign political dictator. A Far Eastern Commission was set up by the three major allies: the U.S., Britain, and the Soviet Union, with authority over all the Pacific allies, to formulate policies and review the General's actions. In practice. however, MacArthur retained the final decision.

Clearly, from his impeccable behavior, he accepted the guidelines proposed in the Potsdam Declaration that concluded the Big Three conference of July-August 1945, which proposed, for Japan as well as for Germany, a postwar program of democracy. The Japanese had been guided towards freedom of speech, of religion, of thought, and of fundamental human rights. Specifically, the Japanese government was ordered to remove all obstacles to the revival and strengthening of democracy.

A basic statement of principles was worked out by the U.S. Departments of State, War, and the Navy. Japan, as envisaged, was no longer a menace to the United States or indeed to world peace and security. On its side, the U.S. undertook not to impose a form of government not supported by the freely expressed will of the Japanese people. Militarists and militarism would be "totally eliminated" from political, economic, and social life. Not only would the Japanese people be encouraged to respect individual freedom and human rights, they would also be given the opportunity to develop an economy adequate to meet their peacetime requirements.

Those were the guiding principles, but the man who turned these dreams into reality was of course Douglas MacArthur. He moved swiftly to remove the props of Japan's militarist state. For a start, the armed services were demobilized. Nationalist organizations were abolished, and Shinto (the state religion) was disestablished. A vast purge of commissioned wartime officers and top executives in industry was instituted. He oversaw the establishment in Tokyo of an international tribunal (much as had happened in defeated Germany) which sentenced General Tojo and six colleagues to death, sixteen to life imprisonment, and two to shorter terms. Tojo and his colleagues were hanged.

The principles of MacArthur's proposed constitutional changes were sound by Western standards but revolutionary to the minds of Japanese leaders. In October 1945, a Japanese Cabinet committee chaired by Matsumoto Joji started work on revising the Meiji Constitution. In February 1946 the Committee presented its new draft Constitution to MacArthur who immediately realized, on reading translated passages, that the changes proposed by the Japanese were too few and too superficial. Under his orders, a new and more radical draft was rushed out and presented to the Japanese. Some of the senior members of the Japanese committee made their objections clear but, perhaps to their surprise, the emperor approved of them.

In April 1946, new elections had been held, in which Japanese women, for the first time, were permitted to vote. The new Diet approved of MacArthur's reforms with only minor amendments, and the new Constitution came into effect on 3 May 1947. In line with MacArthur's reforms the Japanese people were committed to ensuring peaceful cooperation with all nations and to the blessing of liberty for themselves and their descendants. A thirty-one-article bill of rights was incorporated into the new Constitution, along with the renunciation of war and a pledge that "land, sea and air forces, as well as other war potential, will never be maintained."

The role of the emperor was, in effect, downgraded. Indeed on 1 January 1946, he had already renounced the traditional claim to divinity. Nevertheless the emperor was described as the "symbol of the State and the unity of the people, deriving his position from the will of the people with whom resides sovereign power." There would be a two-chamber Diet, with decisive power in the hands of a House of Representatives. The former House of Peers yielded to a House of Councillors, with six-year terms for members. The Diet would choose the prime minister from its own members. There would also be an independent judiciary, with the right to review Constitutional provisions.

Thus, at almost the same time, at two opposite ends of the earth, two destructive dictatorships had been defeated and turned into non-aggressive democracies

Korean Footnote

At the end of World War II, Korea was divided into two states along the 38th parallel, roughly equal in size: a Communist state in the north, a potentially democratic one in the south. With Stalin"s personal approval, the North Korean leader, Kim Il Sung, launched a military invasion of the south on 25 June 1950, crossing the 38th parallel at eleven points. A few days later, President Harry Truman appointed General MacArthur commander-in-chief of the United Nations forces in South Korea.

By 5 September, the North Korean invaders controlled most of the peninsula. Eight days later, MacArthur's forces made an amphibious landing at Inchon which took the invaders by surprise. On 24 November, he ordered a general offensive to end the war. His plans were frustrated when substantial Chinese forces intervened, taking the Americans by surprise. By late November, the UN forces had been driven back to the vicinity of the 38th parallel.

MacArthur was not, by nature, a commander ready to settle for anything less than a complete defeat and retreat of the invaders, Chinese and Korean. On 14 March 1951 the UN forces had reoccupied Seoul, the southern capital, with the invaders in full retreat. Ten days later, MacArthur announced his readiness to meet the commanders of the North Korean and Chinese forces to discuss a peace settlement. Five days later Beijing rejected his offer.

An unexpected career reverse then hit General MacArthur, without any prior warning. On 10 April, he was sacked by President Truman and replaced by General Matthew R. Ridgway.[2]

Why did the president dismiss the greatest American soldier, who had done more than anyone else not only to defeat the Japanese, but also to guide the vanquished along the lines of peaceful democracy? Although the full reasons have never, to my knowledge, been officially revealed, it is known that there was indeed a reason of major importance. The general had worked out a plan: to attack China itself with atomic weapons and force the Chinese to withdraw their troops from Korea..

Truman himself had considered using the atomic bomb. Indeed, at a press conference on 30 November, he had allowed himself to be trapped into a statement of which he soon repented. He had told the assembled reporters that the U.S. would "take whatever steps are necessary to meet the military situation." A hand was raise and a voice was heard, asking (not surprisingly, at that time):"Will that include the atomic bomb?"

The president's response was ill-judged: "That includes every weapon we have."

"Mr President, you said 'every weapon we have'. Does that mean that there is active consideration of the use of the atomic bomb?"

"There has always been active consideration of its use."

The Joint Chiefs of Staff had already recommended against the use of the atomic bomb against North Korea, but the, president who, constitutionally, was supreme commander, carried more weight than they did. His unrehearsed and ill-thought out statement caused consternation in the West. In a visit to Washington in late November 1950, Prime Minister Clement Attlee had opposed any use of the atomic bomb in the Korean War. At issue was the avoidance of any major Western attack on China which, it was felt, would escalate the limited war in Korea into a third World War, since Russia would be bound to intervene on China's side. In view of MacArthur's insistence that the atomic bomb be used, the president had no alternative but to accept MacArthur's resignation.

Part 2

Post-Colonial Conflicts

5

France's Defeat in Vietnam (1945-1954)

Two of the most significant wars of the twentieth century, in the context of this book, involved France's distant colony: Vietnam. The first was a war of independence, launched by the Vietnamese Communist Party, controlled by its leader, Ho Chi Minh. It did indeed bring independence to Vietnam, but not democracy. The French colonial forces had been defeated; but only the northern half of the country came under Ho's communist rule. The second Vietnam war was also launched by Ho Chi Minh, with the aim of extending communist rule over the southern half of the country. In this it succeeded, in that the U.S. forces were defeated and compelled to pull out.

Neither war had been started by a Western democratic power. In each case therefore, military defeat automatically brought political defeat to a Western democracy. The political consequences of both wars are only similar in some respects; in others, they differ significantly. For this reason, I deal with them in two linked, but separate chapters.

The victor in each war was Ho Chi Minh, and his Communist Party of Indochina. His original name was Nguyen Tat Thanh. During his early secret life as a revolutionary, he called himself Nguyen Ai Quoc, meaning "Nguyen the Patriot." In his later years of fame, he switched to Ho Chi Minh, which meant "the Enlightened One." By any measure, he had an extraordinary career. Aged nineteen, he traveled to Europe as a cabin boy on a merchant ship. He worked in London as assistant to French chef Escoffier, In May 1920, he attended a conference of the French Socialist Party (which he had joined) in Tours. There, the party split into two, when a breakaway wing called itself the French Communist Party, with Ho as a founder member. He went on to Moscow in 1923 to study Marxism-Leninism in the "University of the Toilers of the East." In 1925, Stalin sent him to South China as a Comintern agent. Five years later, he had founded the Communist Party of Indochina—an interesting choice of name, as "Indochina" included the adjacent French colonies of Laos and Cambodia, as well as his own country of birth, Vietnam: the most important of the three.[1]

It is fair, and important, to add that although Ho was a committed Moscow-line Communist, his guiding principle was patriotism. He chose communism as the best road to independence for Vietnam, and stuck to it for the whole of his adult life.

Politically, he was the unrivalled Vietnamese leader. Militarily, his unrivalled comrade was Vo Nguyen Giap, who attended the congress of May 1941 convened by Ho in China's Kwangsi province, sixty miles north of the Vietnam border. The outcome was the League for the Independence of Vietnam, known as the Vietminh from two syllables of its full name in Vietnamese: Vietnam Doc Lap Dong Minh Hoi—a characteristic communist-nationalist front.

By 1943, Giap had 10,000 men under arms ready to fight the French colonialists out of their country. In later years, he emerged as a military genius, capable of defeating the French armed forces.

World War II, as is well known, marked the end, or the beginning of the end, of West European colonialism. In the context of France's Indochinese colonies, the context was complicated by the Japanese decision to seize control of them from France's Vichy authorities—an inadequate projection, thousands of miles from France of Marshal Pétain's collaborationist government with its Nazi conquerors.

On 9 March 1945, the Japanese seized control of the whole of Indochina after overwhelming all the French garrisons. That day, they arrested the hereditary Emperor of Vietnam, Bao Dai, and ordered him to proclaim the independence of Vietnam and form a government. In no position to argue, Bao Dai did as he was told, but not surprisingly the government he proclaimed was not recognized by any of the elements that mattered: the Vichy government, the Communists or the nationalists. One of the potential nationalist leaders who refused was Ngo Dinh Diem, at that time a little known politician, but soon to become famous. He had served as minister of the interior in an earlier Bao Dai government under French control.

With typical skill. Ho Chi Minh seized his moment on 7 August—the day after the U.S. Air Force had dropped the world's first atomic bomb on Hiroshima. As a major step towards communist-controlled independence, Ho set up a Vietnamese People's Liberation Committee, with himself as chairman. A few days later, Bao Dai took up Ho's challenge in his own, less defiant, manner. He addressed an appeal to France's new leader, General de Gaulle. His form of words was carefully chosen to appeal to the General: "You have suffered too much during four deadly years not to understand that the Vietnamese people, who have a history of twenty centuries and an often glorious past, no longer wish, can no longer support, any foreign domination."[2]

Full of his own importance, and with little personal interest in Indochina's independence, de Gaulle simply ignored Bao Dai's appeal. Ho Chi Minh, however, saw Bao Dai as a potential instrument to further his own ambitions. His representatives persuaded the Emperor to abdicate, hand over the Imperial Seals and join him in Hanoi as his "Supreme Adviser."

On 28 August, Ho dissolved his Liberation Committee and set up a second provisional government. Five days later, on 2 September, he proclaimed Vietnam's independence, in a carefully worded document that made no mention of the Communist Manifesto but invoked both the American Declaration of Independence and France's Declaration of the Rights of Man. In the confused situation that followed, no fewer than four foreign powers were involved: China, Britain, France and the United States.

Taking advantage of the Potsdam conference of 1945, from which France had been excluded, China started a military occupation of Indochina north of the 16th parallel. It became clear to observers that it suited Ho better to deal with Vietnam's ancient Chinese colonizers than with the French "newcomers."

On the French side, Jean Sainteny, a young Resistance leader, came in from China where he had headed a French intelligence network known as M5. Not surprisingly, in the context of the times, he soon clashed with the American Office of Strategic Services (OSS, precursor of the CIA). More surprisingly, the OSS tried to offset Sainteny by making approaches to the Vietminh. Meanwhile, Sainteny and his friends took possession of the deserted French governor-general's palace. and, rewardingly, Giap met him on 27 August.

Enter the fourth protagonists: the British. Again, under the Potsdam agreements, the British planned to occupy southern Indochina; which they did, under Major-General Douglas Gracey's British-Indian force who landed in Saigon in the first week of September. The rival French force, under the popular General Philippe Leclerc, hero of his country's liberation, did not start landing until 3 October. The British had no intention of seizing even a tiny portion of Indochina for themselves, and indeed started freeing and arming French prisoners-of-war shortly their arrival. This was not the kind of news the Vietminh wanted, and in retaliation they massacred French civilians and set up networks of Vietminh guerrillas.

It is interesting, in retrospect, to recall France's plans after recuperating its territories from the Japanese. Freedom for its colonies was simply not on the agenda. The French were ready to extend the rights of colonized peoples to the levels already enjoyed French citizens (freedom of speech, equality of rights) with citizens of the French Republic Such reforms were to be made available to the inhabitants of the freed French colonies. But *not* full independence. In a declaration by General de Gaulle's provisional government on 24 March 1945, the colonial peoples were indeed offered "freedom" and "economic autonomy" but only within a French Union. Full independence was unthinkable.

Not surprisingly, Ho Chi Minh, although aware of France's attitude, hoped to argue the case for independence convincingly enough to change France's collective mind on this issue. He tried twice: first in talks in Hanoi and Dalat. and later in France, at Fontainebleau. France's commitment to offer free membership of the French Union remained immoveable as the last round of talks ended in September 1946 with nothing to show.

This negative ending was taken by Ho Chi Minh as an invitation to war. On 19 December that year, Giap launched a general offensive against the French forces in Vietnam's northern province of Tonking. Having failed with Ho, the French turned to Bao Dai, who had broken with Ho on realizing the communist character of the provisional government, and after prolonged negotiations, in which the French offered no more to the ex-emperor than to Ho Chi Minh. Nevertheless. Bao Dai formally accepted the French offer of "independence within the French Union" (au sein de l'Union Française) and signed the proposed agreements, even though they recognized France's right to maintain military bases in the country.

The War

It took longer than Giap must have hoped to demonstrate his military genius, for initially, he was handicapped by inadequate armaments. Although he had boosted the number of trained men from 10,000 to 60,000, only two out of three had rifles. Not surprisingly, then, in the early stages, the French drove the Vietminh forces back to the mountains on the Chinese border. There, Giap bided his time.

His patience was rewarded, three years later, when Mao Tse-tung's victorious forces reached the North Vietnam border. From then on, Giap could rely on a virtually limitless supply of arms and, if things went wrong, a cross-border sanctuary . But Giap's time-biding went on: in the north, not in the south, where in large numbers, Vietminh terrorists seized control of the villages ... by terrorizing the villagers.

On 1 October 1950, nearly a year after the arrival of the Chinese forces along the border, Giap launched his first offensive. By then, his clandestine forces had gained control over a 100-mile strip of jungle south of the French-held frontier forts. By the 17th, their garrisons had been driven out of all the border forts: Lao Kay, Ha Giang, Cao Bang, Lang Son and Mon Kay. Giap's victory ranks as "the greatest defeat in the history of French colonial warfare."[3] A further period of Vietminh preparations followed.

The French, meanwhile, with Bao Dai's support, opened an inter-service training college at Dalat. In return for the ex-emperor's support, the French had offered him the supreme command of the newly created Vietnamese Army, which of course he accepted. The French, however, would retain the ultimate Supreme Command, which was entrusted to General de Lattre de Tassigny, a brilliant if temperamental professional who had distinguished himself in the latter phase of World War II by liberating Eastern France from the Mediterranean to the Vosges mountains. The General arrived in Saigon on 17 December, and created an immediate sensation by the victorious tone of his speeches.

He confirmed his reputation in a major clash between the two armies on 14 January at Vinh Yen, thirty miles from Hanoi, Giap's target. The Red Delta Battle, as it became known, lasted four months and ended in a major French victory, after countering Giap's human waves with napalm and high explosives. The defeated Vietminh force had lost 6,000 killed and 500 prisoners. De Lattre's pride in victory was short-lived, for shortly after it he fell ill with cancer, and was flown to Paris, where he died in hospital on 11 January 1952.[4]

The flamboyant de Lattre's successor was the colorless Henri Navarre, a virtually unknown general, for the best of reasons: he had made his career in military intelligence. Arriving in Indochina in May 1953, he was appalled to learn that, despite the presence of more than 100,000 French Union troops in the Red River Delta, the Vietminh controlled 5,000 of the 7,000 villages in that area.

His plan of action was to force the Vietminh into a major military clash, as distinct from the minor but depressing clashes of guerrilla war, usually won by the elusive guerrillas. His choice of a battle-ground made sense only to himself: Dien Bien Phu was a few miles from the Laotian border on the Tonking side, but some 200 miles north of Hanoi. Apparently unable to project his mind into the enemy reality, he had persuaded himself that the Vietminh would be unable to maintain more than two divisions and 20,000 coolies in that area. He was aware that the village was surrounded by hills and therefore vulnerable to artillery fire, but this point left him unworried as he comforted himself with the thought that the Vietminh had virtually no artillery.

In the event, some 80,000 coolies, wheeling bicycles with arms and food through the jungle, were able to keep four Vietminh divisions supplied with arms and food. The arms included a substantial supply of heavy artillery, most of it of American origin, culled by the Chinese Communists in their victorious civil war against Marshal Chiang Kai-shek's U.S.-supported Nationalists. On his side, Navarre had a considerably smaller and very mixed force, in which French troops were outnumbered by a combination of Vietnamese recruits, Foreign Legionnaires, and one-fifth North Africans, mainly from Morocco.

The siege of Dien Bien Phu ended on the night of 6 May 1954, signaling the end of the first Indochina war.

The Truncated Peace

Several competing political forces were represented at the Geneva conference in the spring of 1954. The French, after their defeat by the Vietnamese Communists, were ready—indeed eager—to surrender their sovereignty over their former colonial territories: Vietnam, Laos and Cambodia, while preserving, as far as possible, a measure of their former economic and cultural advantages, not least the French language and culture. The North Vietnamese, led by the revolutionary Ho Chi Minh, hoped to take over the whole of their formerly colonized country, but were ready, if pressed, to make do with the northern half, but only on a temporary basis. In these fundamental aims, they were of course supported by both the major communist powers: the Soviet Union and Communist China.

The South Vietnamese dreaded, above all else, to be taken over by the North Vietnamese Communists. They were well aware that the only Western power capable of coping with a Communist challenge from the north was of course the United States. Their aim, therefore, was to seek to maintain U.S. support, both economically and militarily.

Finally, the U.S., which had made it possible for South Vietnam to survive the French defeat at Dien Bien Phu, was prepared to defend South Vietnam against international and internal Communist threats. Politically, though not militarily, the United Kingdom supported the U.S. stand.

Ho Chi Minh and his communist government, as the clear victors in this first Vietnam War, had grounds to claim that the whole of Vietnam should now be brought under communist rule. However, despite support from the two ideological giants—the Soviet Union and the Chinese People's Democratic Republic—and France's defeat, he was confronted not only by defeated France but by the unchallenged leader of the Western world: the United States of America, with support from its former imperial sovereign: the United Kingdom.

The leader of the north Vietnamese delegation, the party ideologue Pham Van Dong, had been instructed to accept the partition of Vietnam, reluctantly, but to press for division along the 13th parallel, which would have given Ho's government control over two-thirds of his country. Instead, the leader of the French delegation, the Radical-Socialist Pierre Mendès-France, persuaded the Soviet foreign minister, Vyacheslav Molotov, to agree to a dividing line along the 17th parallel, to bring the conference to a relatively harmonious end. On gaining power in Paris a month earlier, he had pledged to end the war within a month, or resign. The clocks in Geneva's Palais des Nations are reported to have been stopped at midnight on 20 July, to enable him to fulfill his pledge without losing the premiership.5

On paper, and in the minds of an ill-informed public, the Geneva settlement looked like a lasting peace. To the better informed, it was merely a temporary truce, paving the way for a second war.

6

The Algerian Disaster (1945-1962)

The history of Western imperialism is packed with contradictions. The spread of Western power—principally of Britain and France—over distant countries, with non-European customs and languages, was originally a form of despotism in that the peoples concerned had not been consulted over the imported methods of government. And yet, the birth and spread of democratic government in the colonial powers themselves inevitably created a desire for it among the colonized peoples.

In this context the colonial history of Algeria—one of France's most important colonies—is of special interest. The Phoenicians had established trading settlements on the Algerians coast, which were taken over by the Romans. The white-skinned Berbers, at that time the main population, were converted to Christianity by the fifth century, but were, in effect, colonized by the barbarian Vandals in the next century. Nor was this all. The Byzantines took over from the Vandals; then came the conquering spread of the Arabs bearing the Prophet's new religion of Islam in the seventh century. For four centuries, Algeria enjoyed prosperity by the standards of that time; which came to an end when the nomad Hillals, from Egypt, moved in. The Hillals disrupted the relatively peaceful Berber kingdoms, by that time flourishing. By the early sixteenth century, the town and harbor of Algiers had come under Turkish control.

Not long after, piracy took over. The pirates, who styled themselves "Corsairs," turned Algiers into their headquarters. The most notorious of the pirate leaders, Horuk Barbarossa, became a redoubtable pest in the eyes of the West's maritime powers: England and Holland, France and Spain, and later America. The closest of the maritime powers was of course France, and in 1830 the French fleet, commanded by Marshal Bertrand Clausel, bombarded the pirate regime into submission. The "submission," however, was far from total. France annexed Algeria and appointed Clausel as its governor. It took fifty-three years for the new colony to be subdued. By that time, Clausel had been dead forty-one years.

In Algeria, as in Vietnam and elsewhere, the advent of communism would dominate the nationalist movement in the 1930s, and on into the 1950s. The first Algerian independence leader of note, Messali Hadj, was a former member of the Communist Party, whose organization styled itself the Étoile Nord-Africaine (North African Star), whose aim was independence and whose tactics included violence. It shifted to a more specific title in 1937: the Parti Populaire Algérien (Algerian People's Party). Perhaps to confuse the colonial power, further name changes followed: Amis du Manifeste [et] de la Liberté, (Friends of the Manifesto and of Freedom) and the Mouvement pour le Triomphe des Libertés Démocratiques (Movement for the Triumph of Democratic Freedoms).

In the event, the key role was enshrined in a definitive name change, which surfaced in 1954: the Front de Libération Nationale or FLN (National Liberation Front), with its military branch, the Armée de Libération Nationale or ALN (National Liberation Army).

The best-known leaders of the FLN were Mohammed Ben Bella, Belkacem Krim and Ferhat Abbas. In the field of revolutionary activism, including violence, the real leader was Ben Bella, who had served in the French forces. A reserve warrant officer on his discharge in 1945, he had won the Médaille Militaire and been mentioned four times in dispatches. On returning to his hometown of Marnia, in Oranie, he had tried orthodox politics and been elected as a municipal councilor of Messali Hadj's legal party, the MTLD. But his flirtation with orthodoxy was short-lived. By 1949, he headed the MTLD's terrorist wing, the Organization Spéciale.[1]

He was forced out of the conflict in a totally unpredictable French intelligence trap. With four other Algerian rebels. he was flying from Morocco's capital, Rabat, to Algeria in October 1956 in a French DC-3 aircraft. While in flight, the French pilot had been ordered, by radioed instructions, to land in Tunis, not Algiers. When the plane landed, Ben Bella was arrested, and imprisoned for six years.

Born in 1924, Belkacem Krim was six years younger than Ben Bella. He had joined Marshal Pétain's collaborationist army in 1942, and served until his discharge three years later, when the war ended. When the rebellion broke out, nine years later (in 1954), he too opted for terrorism, with his moderate compatriots as victims, as well as the French.[2]

Although no less dedicated to Algerian independence, Ferhat Abbas was not involved in terrorism Born in 1899, he was seventeen years older than Ben Bella, and twenty-three years older than Belkacem Krim. He attended the French Lycée in Philippeville and graduated in pharmacy at the University of Algiers. In his student days and during the war, he had been unshakably pro-French, aiming at legal equality of Moslem and European Algerians as the key to transforming Algeria into a province of France. During the war, he had supported Pétain, praising France in a report to the marshal as "the nation of generous traditions which, for generations has sacrificed herself for the respect of the individual and human dignity in the world."

He went on, however, to tell Pétain that all Algerians would continue to agitate until all French privileges were abolished and all the people in Algeria were equal before the law. By then, frustration had got the better of his pacifism. On 31 March 1943, after the Allied landings in North Africa, he presented a manifesto he had drafted to the governor-general of Algeria, Peyrouton. It called for an Algerian constitution guaranteeing the freedom and equality of all inhabitants, regardless of race or religion, along with the recognition of Arabic as an official language, agrarian reform, freedom of the press and association, and compulsory education for both sexes, plus freedom of worship.

To further these evolutionary—not revolutionary—ideas, Ferhat Abbas founded the Union du Peuple Algérien, the controversial core of which was a passage on the "personal Koranic status" of the Algerian Muslims. At this point, Abbas did indeed get to the heart of the problem. Not unnaturally, he wanted his fellow-Muslims to enjoy the rights of their co-religionists as in constitutionally Muslim States, such as Saudi Arabia. In other words, French citizens of the Muslim faith should have the right to practice Islamic customs, not least that of polygamy.

The prevailing view of the French legal establishment at that time was that since a personal Koranic status was incompatible with the rights and obligations of French citizenship the latter must remain unavailable to Alegerian Muslims unless they renounced the said Koranic status. This view prevailed officially, until 11 December 1943 when General de Gaulle's French National Liberation Committee, sitting in Algiers decided to confer French citizenship on the Muslim "elites" immediately while they retained their Koranic status. A few weeks later, on 7 March 1944, an official Ordinance, promulgated by de Gaulle personally, gave the Committee's ordinance the force of law.

In the eyes of the Muslim majority, this decision met their needs, and indeed it did; and therefore attracted fierce and resentful objections from the European French of Algeria, who saw it as bound to pave the way to political domination by the Muslims. Not unnaturally, the European settlers regarded the principle of the "double electoral college" as the guarantee of their privileged position. The European "college" held the levers of power in that its existence prevented the Muslim community from gaining a majority in all the elected bodies. Clearly, the 1944 Ordinance opened the door to Muslim domination of a single electoral college.

Despite this controversial forward step, the French stuck to the 1944 Ordinance and indeed reaffirmed it on 20 September 1947 in a new Statute of Algeria. By this time, de Gaulle was out of power, having resigned in January because of continued opposition from Leftist deputies. The new and elaborate Algerian document went so far as to meet many of the demands of Ferhat Abbas's manifesto. As if to emphasize the point, Abbas was elected to the Algerian Assembly, which was still under the dual college.

Despite this visible progress, it was ever clearer to Abbas and his followers that the French settlers were determined to retain their privileges and their political ascendancy. Both the general election of 1947 and the partial election of 1951 were faked in their favor. By then, Ben Bella's and Belkacem Krim's preparations for their armed rebellion were far advanced. Even when the fighting and the terror broke out, Abbas clung—though not for much longer—to his receding hopes for a peaceful solution.

As I noted in my 1960 work, *The Rebels*,4 Ferhat Abbas's final disillusionment bears a date: 6 February 1956. On that day the new French prime minister, Guy Mollet, visited Algiers, where he was greeted by slogan shouting and a shower of rotten tomatoes. Mollet was a socialist and had embraced a policy of reforms leading to Algeria's full independence. The rotten tomatoes changed his views: by the time he returned to Paris, he had decided on a new policy of pure repression.

This major change of direction marked the inevitable turning point for Ferhat Abbas. He dissolved his own UDMA, flew to Cairo, and announced: "There is only the FLN." This decisive change of course brought him, for the first time, in line with his more extreme allies, Ben Bella (who would soon be abducted by the French) and Belkacem Krim. The war was on.

It was, in both senses, a "dirty" war: a war of terrorism on the rebel Algerian side, and terror on the French side. It is important to grasp that for years the French as a whole had thought of Algeria as simply an extension of their own country: a prolongation of their *patrie* beyond the Mediterranean; whereas they recognized the two neighboring North African countries—Tunisia and Morocco—as colonies. The importance of the distinction is both intellectual and emotive: colonies may achieve independence; an overseas extension of the motherland already has it (intellectually speaking) and therefore does not have to fight for it.

The Algerian war would prove them wrong, but not before a dramatic, but in the end a failed, military attempt to seize power.

The circumstances of the war need to be restated and understood. As we have recalled in the preceding chapter, the prolonged first Indochina war ended with the crushing French defeat at Dien Bien Phu in May 1954. Each of the six years of war had killed off the entire current promotion of the French Military Academy at Saint-Cyr. Eight years of unavailing combat had climaxed in a matching diplomatic defeat in the international peace talks that followed at Geneva.

Demoralized, the French armed forces found a civilian scapegoat in the corrupt politicians (as the military men saw it) in Paris. In a matter of months, the veteran defeated officers of the Indochina war found themselves, in that prolongation of the situation: an impotent country fighting an apparently similar war against a guerrilla army supported by the FLN, a terrorist organization not dissimilar to the Vietminh in Indochina.

Despite their defeat in Indochina, the Army officers felt more optimistic this time. For one thing, in Vietnam they had faced the severe geographical handicap of fighting a war 8,000 miles from the fatherland, whereas Algeria, that "prolongation of France" was merely across that seawater "lake" called the Mediterranean. Nor was there a hostile great power behind the terrorists and guerrillas, whereas China and the Soviet Union had loomed behind the Vietminh in Indochina.

Also in France's favor, as the Army saw it, was the fact that Algeria was part of France. In their eyes, therefore, anybody who opposed French Algeria must be a traitor. At this time, there was no shortage of French writers, academics and deputies who were advocating independence for Algeria, as the only sensible way of putting an end to an unnecessary war, as they saw it. In military eyes, such people were traitors to be pilloried and even (as happened) to be "plasticated" with homemade explosives.[5]

Among the prominent Frenchmen who were critical of the Army's role in preventing the advent of independence for Algeria were two in particular: the former French prime minister, Mendès-France, who had "given away" Indochina and resigned from the Guy Mollet government when it had opted for a military crushing of the Algerian uprising; and the leading sociologist Raymond Aron who had advocated abandoning Algeria in his best-selling work *La Tragédie algérienne*. As it happened, both Mendès-France and Raymond Aron were Jews, which led to a wave of anti-Semitism in the armed forces.

Political Reflections

The perspective of several decades does not encourage contentment about the Algerian war.

First: could it have been avoided? The clear answer is yes. The widely accepted view that Algeria was an extension of France was demonstrably unfounded, since its Constitution, unlike that of mainland France, was not democratic. Racially and culturally, it was clearly a colony, since the dominant racial group (European and basically Christian) was a minority. The ethnic majority, Muslim in culture and religion, consisting mainly of Arabs or Berbers, was kept in its lower political place through the dual electoral college, which guaranteed French power.

The hope aroused by the new French premier, the socialist Guy Mollet, in 1954, had been reversed by that shower of rotten tomatoes. On the spot, the ultimate political power rested in military hands, and one thought dominated military thinking: the thirst for a victory after the humiliation of defeat in Indochina. Initially, the military looked upon General de Gaulle as its designated savior: the man with a modest military career who had achieved full political power in the referendum and election that had approved his new Constitution for a Fifth French Republic and elected him president in January 1959 with wide-ranging powers.

De Gaulle, however, was less interested in French control over Algeria than the military leaders had assumed. His interests were wider: French domination of the European movement, pulling France out of NATO, keeping Britain out of Europe, improving France's relations with the Soviet Union.*

Already, in a speech to the nation on 4 November 1960, de Gaulle had used the term "Algerian Republic," for the first time. On the 16th, he announced that a referendum on Algerian self-determination would be held on 8 January 1961. In the event, 76 percent of the 27 million voters in France and Algeria went to the polls. The same percentage—76—of them in France voted in favor, and in Algeria itself 70 percent, for independence.

Belatedly realizing the turn of events, four French Army Generals (Challe, Salan, Jouhaud, and Zeller) jointly announced in Algiers that they had taken over control of Algeria. De Gaulle's response was immediate and powerful. He proclaimed a state of emergency, suspended all air and sea transport to Algeria and took over emergency powers under Article 16 of his new Constitution.

In March and April 1961, all four of the rebellious generals were arrested and sentenced to life imprisonment. Algeria went on to full independence in July 1962. But to those who had fought the French to free their country from foreign domination, independence did not mean political freedom in the Western sense. It certainly did not equate with freedom in the eyes of its first president, Ahmed Ben Bella, now released from his controversial imprisonment.

While it lasted, effective authority, at least in Algiers, had been grasped by a rival nationalist, Ben Youssef Ben Khedda, a former secretary-general of the Mouvement pour le Triomphe des Libertés Démocratiques (MTKD or Movement for the Triumph of Democratic Freedoms). To counter Ben Khedda, Ben Bella and a fellow ex-prisoner, also just released, Colonel Houari Boumedienne, set up a Bureau Politique: not in Algiers, which they did not yet control, but in Tlemcen, near the Moroccan border. In September, Ben Bella's ALN forces advanced on Algeria. A pharmacist by profession, Ben Khedda did not consider himself a military leader, and did nothing to resist his rival. When Ben Bella reached Algiers, he simply took over and purged Ben Khedda and his followers from a single list of candidates to a proposed Constituent Assembly.

If any doubts remained about Ben Bella's political intentions, he removed them by extending his purge to the Communists, the

* I have dealt with these issues in detail in my biographical history, *De Gaulle* (1973). Reissued in a de luxe edition by the Easton Press (Norwalk, Connecticut, 1990) with a foreword by President Richard Nixon. BC.

Messalists and left-wing Socialists. Under Ben Bella's direction, a new constitution, created by Ben Bella, was approved in September 1963 by the National Assembly, consisting mainly of Ben Bella's appointees.

On paper, then, Ben Bella had become in effect Algeria's dictator, sheltering partially behind his new Constitution. His spell of power, however, lasted less than three years. In June 1965, his fellow revolutionary, Boumedienne, overthrew Ben Bella's government in a military coup. Ben Bella spent the next fifteen years under arrest or in exile

Boumedienne died in 1978. There were elections, but only within the party. Constitutionally, post-French Algeria was closer to the Soviet model than to that of France, the former colonial power. In its first congress for more than fifteen years, the FLN elected a Central Committee of 160 members. In turn, the Central Committee elected a secretary-general, Colonel Chadli Benjedid. Echoes of Moscow: Benjedid selected a Political Bureau (Politburo?) of seventeen members, whose nomination was referred to the Central Committee, which confirmed the secretary-general's choice. By FLN's own law, the secretary-general was automatically head of state. So, in effect, Benjedid's ultimate (though unspecified) title was head of state.[6]

In June 1980, the FLN Congress gave "full power" to Bendjedid, who reduced the membership of the Political Bureau from seventeen to seven and awarded himself most of its powers. Despite its significant similarities to the Soviet system, the FLN's guiding ideology is best described as "Islamic socialism." In 1997, Arabic became the language of state schools. Significantly, protests followed as many students felt marginalized in higher studies, which were still conducted in French. The Berbers also protested, since their own language in no way resembled Arabic. In May 1980, Benjedid yielded, at least partially, to Berber protests, by launching Berber language radio programs and university chairs.

The relative peace of independence unfortunately harbored the seeds of unrest and competitive violence. There were three separate ideologies in a potentially deadly competition: Islam, the political Left and the Army. Moreover, the Islamic force was soon split into separate organizations competing for fundamental truth and absolute power.

Warning signs came with extensive riots in the autumn of 1988. On 19 September, President Benjedid, in a broadcast speech, called on Algerians to work harder to reach economic goals. He threatened to sack "corrupt and irresponsible officials." The economic background was not helpful. Oil prices had been falling, and consumer subsidies had been stopped, causing a sharp rise in shopping prices and rising unemployment. Severe food shortages completed the dismal picture.

A series of industrial strikes worsened the picture. Then came the fateful evening of 4 October, when youths in Bab al-Oued—a poor district of Algiers—set fire to police cars and government buildings. Next day gangs of teenagers set fire to buses and cars, attacked state-owned shops and FLN offices, and erected barricades. Borrowing an ideological term from anti-Israeli Arabs in Palestine, the demonstrators started using the term *intifada* (holy war). The police aimed tear gas at the rioting youngsters, without effect. Two days on, the government declared a state of siege and placed the capital under military command. Army reinforcements were brought in, with support from armored cars and tanks.

The following day was Friday, Sabbath of the Islamic world. Clerical leaders called for peaceful protests, while encouraging further rioting elsewhere in the country. In Algiers itself, over the next few days, soldiers fired on crowds with automatic weapons.

The official casualty figures were high, at 161 killed and only 154 injured; but hospital sources put the number of dead at around 600, attributing the higher figure to the reluctance of the injured to go to hospital for fear of being arrested. The official figure of arrests was 3,743.

A period of calm followed, and in July 1989 a law was passed giving opposition parties the right to register legally. On 12 September, three parties—illegal until then—were legalized: the Parti d'Avant-garde Socialiste (the Party of the Socialist Vanguard), the Rassemblement pour la Culture et la Démocratie (Rally for Culture and Democracy) and (the most important of the three), the Front Islamique du Salut (FIS or Islamic Salvation Front). The reason for the importance of the FIS is that its very existence, with a claimed membership of 3 million, challenged the established FLN. A sign of its Islamic extremism came in July when FIS militants burned down the house of a divorced woman "to purge her sins" The "sinning" woman survived, but not her three-year old son.

By early 1990, the number of parties had reached twenty, the latest of which was ex-President Ben Bella's Mouvement pour la démocratie en Algérie. The ex-president was still alive, but in exile in Switzerland, with charges against him still outstanding.

In March1989 all Army representatives withdrew from the ruling FLN's central committee. The reason given was to allow the Army to "devote itself completely" to its duties under the new Constitution. Subsequent events would indicate that the Army was in reality more interested in preparing itself to crush the Islamic fundamentalists in a civil war that loomed ahead.

Meanwhile, normal political events took place as though no threat existed. In a television broadcast on 3 April 1991, President Chadli Benjedid named 27 June as the date for Algeria's first multi-party general election. He had spoken too soon. Following hostile Islamic demonstrations in Algiers and the resignation of Prime Minister Moulloud Hamrouche on 4 June, the election was postponed. That day police fired on stone-throwing demonstrators with rubber bullets and targeted them with tear gas, sending hundreds to hospital. Benjedid declared a state of siege and gave the military emergency powers for the next four months.

Between 30 June and 2 July, the Army occupied the Algiers headquarters of the fundamentalist FIS, closed two mosques and made 700 arrests. On 22 September, the president's office announced that the state of siege would be lifted at the end of the month. The postponed general election would take place on 26 December.

It did, and resulted in a sensational win for the extremist FIS, with an outright winner in 181 of the 231 seats. This victorious result suggested that the FIS would sweep the board when the second round took place, on 16 January 1992.

The second round, however, did not take place as planned. Two days earlier, on 14 January, the Army-controlled High Committee of State (HCS) assumed presidential powers and began to detain FIS activists. Protests and clashes followed in various places. By 10 February, according to official figures, fifty deaths and between 100 and 200 injured were announced. Not surprisingly, the FIS had different figures to announce: 150 dead and 700 injured.

Politically, the situation had changed dramatically. On 11 January, the moderate president, Chadli Benjedid, had announced his resignation. On the 16th, Mohammed Boudiaf returned after twenty-eight years in exile and stepped into the office that awaited him: Chairman of the High Committee of State. On 9 February, the HCS declared a twelve-months state of emergency. The authority conferred upon Boudiaf, however, did not guarantee his well-being. On 29 June, aged seventy-three, he was assassinated, while addressing an audience at the Arts and Culture Club in the East Algerian town of Annaba,

Back to the politics of power. On 4 March, by order of the Algiers Court of Appeal, the fundamentalist FIS was dissolved. In no way, however, did this put an end to the developing civil war. It was not a war between two opposing armies: it was a war between the terrorists of the dissolved FIS and the terror of the de facto military government.

On 26 August 1992 a bomb exploded at Algiers airport, killing nine and injuring 128. No organization claimed responsibility for the explosion. Was the FIS responsible? Two days later, its newssheet, Mihrab al-Jumaa, denied responsibility, declaring that "the Islamic combatants do not authorize any strike without having precisely defined their objective within respect for the rules of Islam." On 1 October, however, four FIS militants (as they were described) were paraded on television, and confessed in detail to the airport attack.

Fundamentalist violence increased dangerously during the next two years; so, too, did illegal violence by the security services: soldiers, police and gendarmes (who constituted the national police).

By the end of 1993, Algeria had become, at least on paper, a military dictatorship. On 27 January 1994 the High Security Council appointed the defense minister, Brigadier General Lamine Zéroual, as Algeria's new president. He was sworn in on the 31st and in a national broadcast that day, he appealed for unity and promised to seek "security, stability and tranquility."

Instead, predictably, the violence increased. To be fair to Zéroual, however, he sought peace talks with the leaders of the banned FIS, but the outcome was yet another resumption of violence. In a statement on 29 October, Zéroual admitted that his efforts at constructive dialogue between the government, the legal political parties and the FIS had failed.

By then much of the Islamic violence came from the Armed Islamic Group (GIA), but the evidence suggested that GIA and the FIS were in fact in league, even though they were reported later to have split.

In September 1988, Zéroual had announced his intention to stand down from office. In a poll held on 15 April 1999, Abdelaziz Bouteflika was elected president as candidate of the FLN and of the military establishment. In statistical terms, his victory was less satisfactory than it might have been, as all the other candidates had jointly announced their decision to boycott the election several days before it took place. There were six of them and they declared that the military had intervened to rig the poll in Bouteflika's favor. On 14 April, the six claimed to have irrefutable evidence that the poll had been corrupted and warned that unless their allegations were investigated, Bouteflika would have no mandate to govern.

Ignoring their declaration, Bouteflika went ahead and the official figures gave him a vote of 7.4 million or 73.8 percent. Protest demonstrations took place in major cities all over Algeria, and criticism came from the French, British and American governments.

Announcing his program after his unsatisfactory victory, Bouteflika declared that deaths in the civil conflict of the past nine years had reached 80,000. There would be no let-up in the military campaign against terrorism. The Islamic Salvation Front (FIS) would not be legalized, but former FIS members would be allowed to join other parties.

On 29 May 1999, Bouteflika made a televized inaugural address. A few days later (on 5 June) the Islamic Salvation Army (AIS), military wing of the banned FIS, announced that it had permanently abandoned its armed struggle against the government, to clear the way for a negotiated settlement of the protracted civil conflict. From that moment, the GIA remained as the major rebel group.

As the violence continued, an increasingly important, though largely passive, role was played by the Berber minority in Algeria. It was a large minority, of around 35 percent. Historically, the Berbers are descendants of the area's original population, particularly numerous in Kabylia, south-west of Algiers, south of Tenes to the east of the capital and in the Aurès mountains of the Sahara desert. In comparison with the Arab majority, the Berbers are more Francophone and their migrants to France exceed those of the Arab Algerians (60 to 40 percent). Their problem is that apart from themselves, no other group in Algeria speaks their language, let alone writes it: their own alphabet, known as *tamazirt* is regarded as archaic. The Berbers call themselves *Imarizen,* which means "free men." As an ethnic minority they are assertive and defiant.[7]

Since the advent of Bouteflika, in particular, every month brought fresh news of ruthless massacres in Berber towns and villages. How many of these were carried out by Arab terrorists, in particular the GIA, and how many by the Army and the gendarmes under different guises is hard to estimate.

Although Bouteflika could boast some successes in restoring order in the cities, failures in the villages were increasing fast. The president of Berber Cultural Movement (MCB), Ferhat Mehenni, put it in these words: "This insurrection was simply the result of a conflict at the summit between Bouteflika and the Army, which has been calling for his departure for the past six months.8 Indeed relations between President Bouteflika and the Army had been deteriorating for some time. For example, the Chief of the General Staff, General Lamari, is not on speaking terms with the president. And so, the tortures and killings go on, with no end in sight. It seems apt to say that no other contemporary conflict has matched the Algerian one for insolubility. Would it be correct to describe it as a war? The question is unanswerable. The original rebellion of the Algerians against French rule certainly qualified as such. But the incessant shedding of largely innocent blood that came twenty years and more after de Gaulle had handed independence to Algeria does not really fall into the normal understanding of what constitutes a war: Islam against the Army (Arab against Arab) could be so described, with the qualification of "Civil." Arabs of the military and civilian categories against the Berbers would be civil war of another kind. Terrorism of the recognized kind (violence for political ends) prevailed but on a confused scale difficult to match elsewhere.

If blame is to be apportioned, it inevitably falls (at the start) upon the French as the occupying colonial power. The rebellious French Army, humiliated by its failure to defeat Ho Chi Minh's anti-colonial uprising in Vietnam, had been determined to keep "French" Algeria in France's hands. General de Gaulle, at least initially, was not against this option, but soon realized that the Algerian problem stood between him and the realization of his major ambition: to play a key role in international affairs. (See above.)

If the French—before, during or even after de Gaulle—had been ready to play a more constructive and protracted role, both the Algerian war and the murderous chaos that followed France's departure as the ultimate authority, could have been avoided. But they were not. Political victory therefore retreated, as the conflict continued, progressing (an oxymoronic term) from relative simplicity to ever increasing complexity.

7

Britain's Malayan "Emergency" (1948-1960)

On 7 September 1945, after the defeat of Japan in World War II, a young Malayan Chinese named Ch'in P'eng, aged twenty-six, took part in the Victory Parade held in London, He was there by official invitation, with a contingent of the Malayan People's Anti-Japanese Army (MPAJA), a communist guerrilla force he had set up. Along with other guerrillas, he had been released from custody when the Japanese began their invasion of the Malayan peninsula. As part of the victory celebration, he was presented with the Order of the British Empire.

Three years later, in June 1948, Ch'in P'eng launched his MPAJA in what turned out to be a prolonged (though in the end unsuccessful) revolt against British colonial rule, known as the Emergency. Why, then, did Ch'in P'eng, OBE, launch this attempt to wrench control of Malaya from the British colonialists and turn it into yet another satellite of the distant Soviet Union?

The origin of Malaya's insurgency is well known: it was plotted in Moscow, where Stalin's right-hand man, Andrei Zhdanov, took on the task of launching a wide range of rebellions designed to expel the "colonial oppressors" from all the colonies in Southeast Asia. To this end, Zhdanov convened representatives of the Western colonies to an international Youth Conference in Calcutta, which met from 19 to 28 February 1948.[1]

The majority of the participants were 900 delegates from various Communist parties or "fellow-traveling" organizations. Those present (by invitation) included a number of non-party "observers." Messages of greeting were read out, from Pandit Nehru of India and his "guru" Mahatma Gandhi (who was assassinated shortly before the conference began), Burma's first independent prime minister, U Nu; and Mrs. Eleanor Roosevelt, widow of the deceased president.

Greetings also came from communist leaders, including Klement Gottwald of Czechoslovakia and Ho Chi Minh of Vietnam. The hidden objective of the Youth Conference was to persuade the communist or pro-communist parties of South-East Asia to launch insurgencies in their respective countries. In the event, rebellions were indeed launched, not only in Malaya, but also in Burma, the Philippines and Indonesia.

Although the Emergency lasted twelve years, it never turned into a major war: mostly, no doubt, because it was "almost entirely conducted inside or on the fringes of some of the thickest jungle to be found in the world."[2]

In Malaya, as distinct from Vietnam, there was a built-in advantage on the government side, in the sense that the guerrillas were overwhelmingly Chinese, while the numerous villages were populated by Malays. Thus the peasantry belonged to the indigenous population, while the enemy could be regarded as intruders from outside. There were, however, many Chinese squatters in the Malayan villages.

On 3 April 1950, a recently retired high-ranking British Army officer, Lieutenant-General Sir Harold Briggs, was brought out of his short retirement to be appointed Director of Operations in Malaya. As Harry Miller points out, the general, with a splendid record in Burma, devised what became known as the "Briggs Plan" for the resettlement of more than half a million Chinese squatters to about 500 new villages.[3]

Nearly four years after the insurgency had broken out, important political changes took place in Britain and, consequently, in Malaya. In Britain, on 25 October 1951, Winston Churchill returned to power after the Conservatives had defeated Clement Attlee's Labour Party. In Malaya, General Gerald Templer took over as high commissioner. After a distinguished war record, he had become director of Military Intelligence in the War Office; a perfect preparation for his role in taking on the communist guerrillas. Intelligent, and strong in character, he was the ideal choice to aim at a total defeat of the enemy. [4]

Templer virtually ensured that outcome in an initial statement entrusted to him by the British government, which opened with words that could be said to have anticipated "political victory," four decades before this book was written: "The policy of the British government is that Malaya should in due course become a fully self-governing nation...."[5]

Towards this enlightened end (in marked contrast to French aims in Indochina) Templer reorganized the police force to enabled one sector of it to return to its normal civilian duties, while the rest would be a para-military contingent operating in the inescapable jungle warfare. He established a single omnipresent intelligence organization; and an information service, geared to psychological warfare, including films and broadcasting.

Having made his important contribution to the Emergency, Templer was recalled to London in 1954, to resume his military career (which is of no further relevance to this chapter).

Templer's successor as High Commissioner, Sir Donald MacGillivray, carried the New Villages a decisively important stage further, by transforming the "old" villages into what became known as "Strategic Hamlets." Relieved of their Chinese minorities, the old villages were overwhelmingly Malay. Thousands of them were provided with defensive fortifications and Home Guard defenders. More, probably, than anything else, this measure turned the Malayan Emergency into a winnable, and in the end a victorious, low-scale war. Having protected the population against subversive and terrorist threats, the British and Malay authorities were able to turn, successfully, to jungle warfare.[6]

In the latter stages of the prolonged Emergency, a new and effective personality played an important role in the final stages of this strangely protracted Emergency: Sir Robert Thompson. An RAF officer during the war, he had escaped from Hong Kong through China and served through the so-called anti-guerrilla operations in Burma. Back in Malaya, where he had served in the Civil Service in 1938, he worked first as deputy secretary and later as secretary for defence concerned exclusively with the Emergency. A patient man, with a methodical approach to the problems faced, he outlined his methods, which he called "Basic Principles of Counter-Insurgency":

- Government measures to restore Authority and regain support from the population in competition with terrorist promises of a better life.
- Governments must function within the law, if necessary introducing new laws to face the terrorist challenge.
- Detention of suspects must be followed by public trials according to the law.
- Coordination between civil and military actions.
- Priority to defeating political subversion, over defeating the guerrillas.
- Securing government base areas against terrorist attacks.[7]

Thompson goes on to stress the importance of intelligence, to penetrate the guerrilla movement and anticipate its plans, and information services to counter subversive propaganda.

By 1952, Ch'in P'eng had evidently lost his confidence, and his hopes of winning the war he had launched four years earlier. He retreated from his headquarters in central Pahang, to a secret retreat in south Thailand. He went on trying, less and less successfully, to run "his" war, then vanished. Nobody saw him or heard from him again.[8]

By then, a promising Malay political leader named Tunku Abdul Rahman had emerged. In July 1955, his Alliance party swept into power in Malaya's first general election, with fifty-one of the fifty-two seats. By then, the Communist guerrillas, deprived of their leader, had dwindled in number to a few hundred. The new prime minister, known as "the Tunku," had attracted Indian and even Chinese community followers. On 31 August 1957, his Alliance government gained another victory, less dominant than the first but with a healthy majority of seventy-four seats out of 104.

It was at this time that the British relinquished the last remains of their power over Malaya. They had pledged independence and lived up to their pledge. Tunku Abdul Rahman became the first elected prime minister of independent Malaya, and for full measure, assumed the portfolios of defense and internal security. Thus, twelve years after Ch'in P'eng's attempt to carry out Zhdanov's orders at the Calcutta Youth Conference, the end of the Emergency was proclaimed.

The price paid had been high, for both sides. The number of guerrillas had declined from about 9,000 to 1,830. Their main shortage was food. By then the troops, police, special constables and Home Guard constituted a force ten times as numerous as the remaining rebels. . The financial cost of the Emergency had been high: a total of some £600 million, for the British and Malayan governments, was hazarded. To quote Harry Miller: "...1,851 members of the security forces had been killed and another 2,526 wounded. Altogether 2,451 civilians had lost their lives. 1,383 had been wounded, and another 807 were missing."

The Communists had suffered more serious casualties—6,398 killed, 1,245 captured, and 1,938 surrendered but unknown statistics were how many of them had died from their wounds, and how many had been executed or had died from natural causes.[9]

On Independence Day, according to Harry Miller, 3,200,000 people were living "free lives in 'white' areas and hundreds of guerrillas were more or less active in central Perak, along the Thai border or in Johore.6 The benefits of independence were numerous. With it, Malaya had become a self-governing member of the Commonwealth. On 12 October, a British-Malayan Treaty of Mutual Assistance and External Defence was signed. The looming problem of Singapore was tackled: was the prosperous island to remain, in effect, an extension of Malaya, or claim its own independence? On 27 May 1957, while the Emergency was still on, an all-party delegation from Singapore, at a meeting with the British colonial authorities, agreed to transform the island into a self-governing State before March 1959. In the event, there was a slight delay: on 3 June 1959, under its new Constitution, Singapore became a self-governing state, and (as a neo-colonial start) the British Governor, Sir William Goode, was appointed "head of state."

Not long after, on 3 December, the first Malayan-born head of state, Inche Yusof bin Ishak, was installed. Two years after victory in the Emergency, on 31 July 1962, an agreement to establish a Federation of Malaysia was signed in London, It comprised not only Malaya and Singapore but also Sarawak, Brunei and British Borneo. Months later, on 16 September, the Federation of Malaysia was formally established. It comprised the areas previously specified, with one exception: Brunei.

However, Singapore had its own independent ambitions. On 7 August 1965, Malaysia agreed to Singapore's withdrawal from the Federation. Two days later, Singapore became independent, and soon after, joined the United Nations. Its first prime minister, Lee Kuan Yew, who stayed in power for thirty-one years, was a highly intelligent and educated man (double first in Law at Cambridge). I first met him in London on a visit, at a time when he was deliberately taciturn to preserve a misleading left-wing image; and later interviewed him for the BBC when he had come to power after shedding his early image. In Singapore as in Malaya, the communist rebellion had been in vain.

In other words, the British in Malaya had achieved political victory, that elusive prize of military wars, which the French had failed to achieve in either Algeria or Vietnam.

8

America's Defeat in Vietnam (1961-1975)

After the Geneva conference of 1954 (see Part II, 1) the main preoccupation of the United States was to protect South Vietnam from a probable North Vietnamese invasion. The Cold War was at its height, and Moscow's policy was to provide North Vietnam with advanced weaponry and support from its worldwide propaganda apparatus, but without involving the Soviet Union directly in a war to further Ho Chi Minh's ambitions.

Ho had set up a subversive apparatus, named the National Front for the Liberation of South Vietnam (NFLSV), usually shortened to the "Viet Cong" as its military sector was called. Ho had opened offices in several countries, including China, Egypt and Cuba; but not, in the early stages, in Moscow. However, a few weeks after the overthrow of the Soviet leader, Nikita Khrushchev, in October 1964, his successor, Leonid Brezhnev, invited the Viet Cong to open an office in Moscow.[1]

On 1 November 1963, the South Vietnamese Premier, Ngo Dinh Diem, had been murdered in a coup by Army generals, impatient with his reluctance to carry out reforms proposed by the Americans. His death caused a pall of uncertainty to descend over President John F. Kennedy, who was himself assassinated a few weeks later. There was no connection between the two events, but important changes of course occurred on both sides of the Cold War.

By the time of Khrushchev's overthrow, several thousand U.S. advisers were stationed in South Vietnam, but no combatant forces. The situation changed dramatically on both sides in 1964-65. On the Communist side, on 26 November 1964, a few weeks after Khrushchev's overthrow, the new leadership in Moscow pledged that they would give North Vietnam all necessary help, including deliveries of modern weapons Ho Chi Minh decided to put Soviet intentions to the test. He chose his moment well. Early in February

1965 the new Soviet premier, Alexei Kosygin, led a powerful Soviet delegation to Hanoi. While it was there, presumably on Ho's orders. Viet Cong guerrillas attacked an American airfield and billet area at Pleiku in the central highlands. The attack was serious: seventy Americans were killed or wounded: three transport planes and seventeen helicopters were destroyed. The American response was immediate: that same day—7 February—U.S. and South Vietnam planes raided North Vietnam targets.

Embarrassed and angry, the Soviet visitors hinted to their North Vietnamese hosts that this would be a good time for peace talks to end hostilities. Their hints fell on deaf ears. The Soviets wanted to avoid a clash with the Americans; the North Vietnamese smelt victory and listened without hearing. Their Chinese friends had converted them to the view that the Americans were "paper tigers." They were on the warpath, and had persuaded themselves that the Americans would not retaliate. In this, they were mistaken.

Having got nowhere, Kosygin and his delegation left Hanoi without any commitment to help. A fortnight later, on 1 March 1965, the rebellious North Vietnamese Lao Dong (Communist Party) boycotted the international Communist conference, which opened, as planned, in Moscow.

On the American side, the new president, Lyndon B. Johnson, decided in June that year to authorize the use of U.S. combat troops in Vietnam. The second Vietnam War had begun.

The new situation suited the new Soviet president, Brezhnev, and his team. They decided, rightly, that shipping military supplies to Hanoi, including new weapons, would not risk a war with the United States. After all, the U.S. clearly remained committed to the view that the U.S. forces were merely trying to support Saigon against the communist rebels, in line with the Geneva conference "solution." A major complication in the problem the Americans had decided to face was the fact that not only were the North Vietnamese being armed by Moscow and Beijing, but they controlled a major and well-organized subversive force in South Vietnam.

The North Vietnamese, still under the command of General Vo Nguyen Giap, would soon put the U.S. intentions to a major, and totally unexpected, test, known as "the Têt offensive." The word "Têt" is of religious, not topographical, significance. In its full version, Têt Nguyen Dàu is the Annamite term for a Chinese religious festival at the end of January. On the 30 January, the Viet Cong forces launched a wide range of attacks on forty-four provincial capitals and six of the largest cities in south Vietnam: thirty-six of the capitals and five of the cities were overrun. The most sensational of the assaults was on the ancient imperial capital of Hué, in central Vietnam, where the attackers stormed the citadel and held on to it for twenty-six days. In the south, the communists even penetrated the grounds of the U.S. embassy in Saigon, although not surprisingly they were routed by Marine guards and Army military police.

General William Westmoreland, the American commander in Vietnam, claimed that by the end of February about 45,000 of the 84,000 attackers had been killed. In military terms, the offensive was a heavy defeat for the Communist forces. In political and psychological terms, however, the winners were the Viet Cong. As the distinguished British authority on guerrilla warfare, Sir Robert Thompson, put it: "...if the offensive resulted in a military defeat within Vietnam, it achieved a striking psychological blow to the United States. It resulted in the abdication of President Johnson, in the opening of 'talks about talks" and in increased dissent. In early November, the president halted all bombing of the north.[2]

The outcome of this presidential decision (apparently intended to boost the chances of Senator Hubert Humphrey, Democratic candidate in the then forthcoming presidential election) failed in that sense, but severely handicapped Johnson's successor in the White House, President Richard Nixon. At a crucial stage of the war, Nixon was unable to order a bombing of the north. He did, however, find a way to bend, though not to break, the rules.

In the early months of his presidency, Nixon sent several secret messages to Hanoi, proposing talks. Ho Chi Minh simply ignored these approaches. Instead, the North Vietnamese launched a second major offensive in February 1969, a few weeks after Nixon's election. Once again, but more so, the attack bore the imprint of General Giap. It came not from North Vietnam, but from South Vietnam's neighbor, Cambodia, where the North Vietnamese Army had joined forces with the Viet Cong.

There was no way Nixon could respond to the new offensive without breaking the ban on U.S. bombing operations. In a sense, it was a worse breach of the ban than if he had ordered the bombing of the north, in that Cambodia had chosen neutrality. Nixon had allowed himself to think, however, that the bombing would weaken North Vietnam's resolve. In this, of course, he was optimistically mistaken.

One of his main concerns was to prevent any publicity for the re-
newed bombing, in breach of the ban imposed by President Johnson.
Indeed, Nixon went so far as to order the U.S. Air Force not to mention the bombing, even in its secret records.[3]

Not unusually, for Nixon, he chose this moment to repatriate 25,000
U.S. troops from Vietnam and to write a friendly and conciliatory
letter to Ho Chi Minh, proposing a phased withdrawal of North Vietnamese forces from South Vietnam, a cessation of hostilities, and
free elections under international supervision. Predictably, Ho's reply was cold and uncompromising. A secret meeting between Secretary of State Henry Kissinger and a North Vietnamese representative
named Mai Van Bo, took place in Paris. Against all the visual evidence, Bo stuck to his official assertion that there were no North
Vietnamese forces in South Vietnam.

This attitude turned out to be Ho Chi Minh's last demonstrated
adherence to Communist falsifications. Not long afterwards, on 3
September 1969, Ho died, aged seventy-nine, thus missing a pre-
dictable North Vietnamese celebration of at least partial victory.

By April 1970, President Nixon had repatriated 115,000 U.S.
troops. On the 20th, he announced the withdrawal of a further
150,000. By then, he had adopted a new principle: that this second
Vietnam war should be "Vietnamized"—in other words, that the U.S.
should provide South Vietnam with the means to carry on the war
until victory.

At that time, Nixon and his administration, including the U.S.
Armed Forces, were up against a phenomenon new to them: a Com-
munist-backed mass demonstration against any further U.S. involvement in the war. The chosen name for the "demo" was the "Vietnam
moratorium." Millions took part in it, not least in the university cam-
puses, where the demonstrations were often violent. There is no doubt
that the constant reporting of American violence (rather than North
Vietnamese) played a major part in the build-up of American public
opinion against the U.S. participation in the war. To give one example, the American massacre of villagers at My Lai in 1968 provoked hostile anti-war demonstrations. In contrast, corresponding
massacres by North Vietnamese forces were not covered by television, thus concentrating hostility against the American forces.

In a televized speech on 3 November 1969, President Nixon declared that the United States would stay in Vietnam until the South
Vietnamese could defend themselves. The truth, however, was that

American aid was reduced, and by 1974 the South Vietnamese forces were cut to about 100,000. By the end of the year, the North Vietnam forces in South Vietnam had reached twenty divisions, with stocks of armor and artillery sufficient to last for a whole year.

As President Nixon's best biographer, Jonathan Aitken, put it: "Vietnam was the curse of the Nixon Presidency. Like a growing tumour it corroded the norms of democracy, debilitated the nation and raised the still unanswered question of whether American public opinion could stomach a long war with heavy casualties in the age of television."[4]

As Aitken rightly added, Nixon was of course aware that he had been elected to end the war, and indeed relished this challenge of peacemaking with excessive optimism; but he failed to see that achieving an acceptable settlement was going to prove "far tougher and longer than he had anticipated. Indeed, before his election, he had told his aides that he did not intend to be "the first American President to lose a war."

Fifteen years later, Nixon was lamenting his country's failure to win. As he put it, in his interesting and enlightening book, *The Real War*:

We had won the war militarily and politically in Vietnam. But defeat was snatched from the jaws of victory because we had lost the war politically in the United States. The peace that was finally won in January 1973 could have been enforced and South Vietnam could be a free nation today. But in a spasm of shortsightedness and spite, the United States threw away what it had gained at such enormous cost.[5]

Further on, Nixon paid me an unexpected compliment, in these words: "The communists had grasped what strategic analyst Brian Crozier said is the central point of revolutionary war: 'that it is won or lost on the home front.'"[6]

Unfortunately, in Western democracies, the home front is not necessarily on the democratic side, even if, in their naivety, they believe themselves to be. As Nixon noted, when the famous film star Jane Fonda went to Hanoi, with the former attorney general, Ramsey Clark, with the commendable intention of seeing for themselves, what they saw (or thought they saw) was typical of communist distortion of the truth: "...they received massive, largely positive American media coverage of their statements praising the treatment of American prisoners of war, who in fact were being subjected to the most barbaric and brutal torture by their North Vietnamese captors."[7]

In January 1975, the North Vietnam forces in South Vietnam launched what proved to be the final offensive against the South, capturing Ban Methuot in the highlands, and on 10 March ringing Saigon with thirteen divisions.

The last few weeks of the war brought no glory to the anti-communist side. On 29 March 1975, the important coastal town of Da Nang, south of Hué, had been captured by the North Vietnamese. For the south, this signaled the end of resistance. Several towns were abandoned without resistance. The port of Qui Nhon, 270 miles northeast of Saigon, was taken over by Communist troops on 1 April, without opposition after thousands of troops and civilians had been evacuated by sea. That same day, Nha Trang, 190 miles northeast of Saigon, was also abandoned. Within the next few days, five other towns were vacated without resistance.

As soldiers left some of these towns without a fight, there was mass looting without opposition, as thousands of terrified civilians tried to escape by sea or air. Two of the towns were reoccupied by southern forces, but not for long. By 19 April the Communists controlled six provinces. From 14 April on, the air base at Bien Hoa, eighteen miles east of Saigon, was shelled by long-distance 180 mm. guns, with strategic benefits, since fighter bombers could no longer use it by night, and half the southern air force had to be evacuated and transferred to Tan Son Nhut air base in Saigon itself.

During the first three weeks of April, the South Vietnamese Army made a stand around Xuan Loc but was forced to evacuate the town, by now ruined and deserted of civilians, on 21 April. In contrast with the terror tactics of the enemy underground fighters, the "official" communist infantry was on good behavior in the closing phases of this last offensive. A French priest told an Agence France-Presse correspondent that there had been no bloodbath when Da Nang was captured. He added that after the violence of the southern troops, the communist forces had restored order by "firm but not brutal" methods.[8]

Although the captured cities were being administered by communist military committees, certain non-communist organizations opposed to the Thieu regime were collaborating with the northern invaders. These included the neutralist National Force of Reconciliation and Concord and various Buddhist bodies. The Catholic clergy were also collaborating with the northern invaders. Indeed, the Archbishop of Saigon, the Most Rev. Paul Nguyen Van Binh, appealed to his co-religionists not to flee from the Communists and to cooperate with the new authorities.

On 30 April, the southern General Duong Van Minh handed over Saigon unconditionally. For a few days, the General himself—the one who had overthrown and murdered Ngo Dinh Diem—found himself in brief power. That day, newsreels captured the sight of North Vietnamese soldiers climbing on to the roof of the American embassy as the last few diplomats boarded helicopters to seek departing refuge in the Seventh Fleet.

Why, and how, did the United States lose the second Vietnam War? The key words in the answer have to be: independence and communism, to which may be added strategic necessity. The U.S. was the leading power in the Western world, the leader of the North Atlantic Treaty Organization, the greatest of all capitalist and democratic powers. The only element missing in its confrontation with North Vietnam was necessity; or indeed sentimental attachment. France, having conquered the three countries of Indochina as part of its eastward colonial expansionism, in competitive rivalry with the United Kingdom, had been ousted from them as a logical consequence of its early defeat by Nazi Germany during World War II. Hence its national urge to restore France's dominance in the face of Ho Chi Minh's rival assertion of power. Nevertheless, it had lost its colonial war and been forced to hand over control to the Vietnamese Communist movement led by Ho Chi Minh, founder member of the French Communist Party, after France's catastrophic defeat at Dien Bien Phu.

Ideally, the U.S. would have welcomed the feat of turning this ancient colony of France, and before that of China for a thousand years, into a democracy. But there was no deep sentimental attachment to its takeover of the military authority from France. It was perhaps surprising that it held on to its role for as long as it did.

On the rival side, from the beginning of his extraordinary career, Ho Chi Minh had equated independence with communism. Having fulfilled the job handed to him by Stalin, he had a built-in incentive to drive the Americans out, as he had driven out the French colonialists. This was the deeper meaning of his long political career. In this permanent aim, he enjoyed the unceasing support of a loyal military genius who did not aspire to political power. Vo Nguyen Giap had demonstrated his extraordinary abilities during the first Indochina War, and preserved them for similar use in the second.

Two other elements played a major role in the two Communist victories: terrorizing control over the peasantry, and support from the leading Communist power, the Soviet Union. Support also came from the new Communist great power: the Chinese People's Republic. This support, however, became considerably greater, because more readily available, in the Second Vietnam War than in the first, when advanced weaponry from Russia was sent to North Vietnam across the vast expanse of China.

The terror element was of a major importance. During the first Vietnam War, and again during the second, the Vietminh, or Viet Cong, had established a reign of total terror over the entire peasantry of South Vietnam. Their typical technique was to burst into a village and murder wives and children before turning onto the husbands who had witnessed the killings. Disemboweling was a normal treatment for wives; hacking off limbs for her children, and emasculation for her husband.[9]

For obvious reasons, it would have been impossible for the French to compete with the Vietnamese Communists in such incidents of scarcely believable horror, even if the desire to compete had existed: for the major reason that only a total, native knowledge of the Vietnamese language and of the local topography would have sufficed. Having seized and maintained total control over the peasantry, the Vietminh and Viet Cong used it to exterminate headmen and notables, doctors and teachers, nurses and agricultural advisers appointed by the central authority in Saigon.

Terror was also the key to indoctrination, aimed especially at the young, with detailed attacks on the existing government, followed by appetizing pledges of a better future for all.

Today, in the early days of the third millennium, Vietnam stands as another example of the frequent political futility of military war: two international wars involving Western democracies, with virtually nothing to show for them. On 3 October 2001, the U.S. Senate ratified by 88 votes to 12 a July 2000 trade agreement between Vietnam and the USA. The expected benefit was to be a wide-ranging cut in U.S. tariffs on Vietnamese exports. The Vietnamese government welcomed the long delayed ratification but strongly criticized the accompanying Vietnamese Human Rights Act, passed by the U.S. House of Representatives, linking non-humanitarian aid to progress in human rights. So much for the delayed legacy of America's war in Vietnam a quarter of a century earlier.

9

The India-Pakistan Antagonism

Two Indian politicians, turned statesmen, created the situation that resulted in an apparently permanent confrontation between two neighboring states; which, many years later, turned the confrontation from a conventional military one into a nuclear one. Their names were Jawaharlal Nehru and Mohammed Ali Jinnah.

India had been by far the largest item in the worldwide British Empire; in population, it was second only to independent China. Although Britain's wartime leader, Winston Churchill, had made it clear that he had no intention of conferring independence upon this huge but disparate land, with complex ethnic divisions and more than 200 languages ("I have not become prime minister to preside over the dissolution of the British Empire.") Having led Britain to victory in World War II, however, his Conservative Party lost the first postwar elections in July 1945, Clement Attlee, the Labour leader who replaced him as prime minister, saw no reason to maintain British sovereignty over India or other British colonies.

At that turning point, two Indian political leaders had to be reckoned with: Gandhi and Nehru. Unlike his disciple Nehru, Mahatma Gandhi was not a politician in the conventional sense. He was an ethical leader, a social reformer and a prophet in revolt against the spread of technology. However, his career ended suddenly when he was assassinated on 30 January 1948.

Although Nehru was one of Gandhi's devoted followers, he was above all a politician, whose guiding ambition was independence for his country. For many years, he was the unchallenged political leader of India's main religious majority: the Hindus. In the late 1920s and early 1930s, however, a challenge came, in the person of Jinnah. After law studies in Britain (notably at Lincoln's Inn Fields), Jinnah had started to carve a legal career for himself in Bombay. A natural outlet for his religious and political instincts had been founded in 1906, under the title of "the All-India Muslim League," but he remained uninvolved in it for seven years. He then joined it, in 1913, after being assured that the League was as devoted to the political emancipation of India as the (mainly Hindu) National Congress that he had joined seven years earlier, in 1906.

During the late 1920s and early 1930s, Jinnah's main political objective was in fact to bring about a permanent rapprochement between India's Hindus and Muslims. At Round-Table Conferences in London between 1930 and 1932, he tabled a Fourteen-Points Plan, which included greater rights for minorities, one-third representation for Muslims in a proposed central legislature and a federal form of government. He pressed the Hindu majority for amendments to Nehru's rival Committee to accept proposals for separate electorates and the reservation of seats for the Muslim minority.

Jinnah's proposals proved unacceptable to the Hindus, while India's Muslims decided to repudiate him as their leader. Disgusted by this turn of events, Jinnah decided to settle in London and practice his profession. In 1935, however, he was persuaded to return to politics. The time for change was indeed ripe. The turning point came independently of his leadership, in late March 1940, when the Muslim League, at a meeting in Lahore, decided to aim at a separate Muslim State, to be known as Pakistan.

This separatist path was of course unacceptable to the Congress and, initially, to the British government. But in the face of Jinnah's skilful tenacity, the Attlee cabinet changed its mind. And so, more reluctantly, did Congress. After negotiations with both sides, the British government announced its new Constitutional plan on 3 June 1947, calling for the partition of Britain's former colony between India and Pakistan. The independence of both countries went into effect on 15 August.

Not surprisingly, the process of partition sparked uncontrolled rioting between Hindus and Muslims, especially in the northern State of Punjab. In the weeks that followed, the combined death toll had reached about 100,000. By late September, nearly two million refugees had been exchanged between the two newly independent Dominions. Already, on 21 September, the new governments had signed a joint statement emphasizing their readiness to eliminate all causes of conflict. The riotous murders did end, but the mutual antagonism between the two Dominions went on, with no end in sight.

One of the main aggravating causes of the mutual antagonism between India and Pakistan is often overlooked: Jawaharlal Nehru, one of the two main leaders (with Gandhi) of Indian independence from British colonial rule, regarded himself as a Kashmiri, although born in the Indian province of Jallahabad. His reason for this view was simply the fact that his ancestors were Kashmiri Brahmins. During the long negotiations on independence between Nehru and the new Viceroy of India, Admiral Earl Mountbatten, Nehru insisted (successfully) on the inclusion of Kashmir into independent India, despite the fact that the State's population was overwhelmingly Muslim, not Hindu. The inclusion went ahead on 26 October. This decision aggravated the Hindu-Muslim tensions, which went on for years. It is indeed fair to say that they have never ended. The following dates are of interest.

- 5 August 1965. The mutual tensions reached a new climax, when India accused Pakistani irregulars of crossing the cease-fire line in Kashmir, and launched a counter-offensive in the direction of Lahore, across the Pakistani border.
- 31 August that year. Pakistan claimed that more than 200 Indian troops had been killed in Kashmir fighting. The Pakistanis asserted that India had sent troops to disputed outposts earlier that month and that Kashmiris loyal to President Ayub Khan were involved in the fighting.[1]

India-Pakistan relations changed dramatically in 1971, after guerrillas in Bangladesh (East Pakistan) launched a rebellion against Pakistani rule in August. By late November some 9 million Bangladeshis had crossed the border, seeking refuge in India. The nucleus of the guerrilla forces consisted of soldiers from the East Bengal Regiment and the East Pakistan Rifles. Apart from local volunteers in increasing numbers, most of the rebels were deserters from the Pakistani occupation forces.

On 4 December 1971 the Indian armed forces launched an integrated ground, air and naval offensive against East Pakistan. On the 16th, the Pakistani forces surrendered unconditionally : the new independent state of Bangladesh was born. It is fair to observe that Jinnah's vision of a new Islamic state consisting of two areas separated by hostile India had seemed doomed to failure from the start.

India lost no time in recognizing the new State of Bangladesh, while not surprisingly, Pakistan itself initiallly declined to do so. The last Indian troops were withdrawn from Bangladesh on 12 March 1972.

The two rival presidents—Mrs. Gandhi of India and Benazir Bhutto of Pakistan—met in Simla between 28 June and 2 July 1972. In effect, the resulting agreement was a mutual recognition of the status quo as it appeared after the clashes in Kashmir and the war in Bangladesh.

If only for geographical reasons, the Pakistanis had not been in a position to deny independence to the formerly Indian territory that became Bangladesh, separated from East Pakistan by the entire northern breadth of India; whereas Kashmir's borders separated Pakistan from neighboring India. They were of course better placed to intervene in the Hindu-controlled Muslim State they coveted. The Simla agreement ended in several paragraphs incorporating important provisions, especially the following:

- Both sides to respect the line of control resulting from the cease-fire of 17 December 1971.
- The withdrawals to be completed within thirty days.

Indian troops would be withdrawn from 5,189 square miles of Pakistani territory in the Punjab and Sind; and Pakistani troops from sixty-nine square miles of Indian territory in the Punjab and Rajasthan. In Kashmir, India would retain 480 square miles of territory west and north of the cease-fire line, and Pakistan fifty-two square miles east of the line. Another way of putting it would be that the late Pandit Nehru's claim to Jammu and Kashmir had been reduced by only 6 percent. In other words, India still retained control of 94 percent of Kashmir's 86,023 square miles.

The Nuclear Confrontation

Let us jump ahead now—sixteen years on to the new world of nuclear confrontation; new, not for the older world of superpowers, but for the newer world of mutually hostile ex-colonies.

15 January 1998. A regional business summit was being held in Bangladesh. On the sidelines were the prime ministers of Pakistan (Nawaz Sharif) and India (I. K. Gujral). It would have made no sense for them not to meet, and indeed it may be presumed that this was the real reason why they were both present. Earlier that month, each side had blamed the other when troops exchanged fire along the de

facto border separating area of Kashmir controlled respectively by each side. The two men had last met in New York in September 1997. Now, despite the exchange of fire, they agreed to continue a dialogue: not in person, but at foreign secretary level.

The sad reality was that both India and Pakistan were preparing to "go nuclear." In fact, India had already done so, twenty-four years earlier. And now, on 11 and 13 May, the Indians detonated five nuclear devices. Would Pakistan respond? Was it capable of doing so? The answer was positive and came within days. On 28 and 30 May, Pakistan exploded six nuclear trial weapons, all underground.

These rival detonations demonstrated to the world that both India and Pakistan had joined the nuclear club, along with the United States, the United Kingdom and France on the Western side, and Russia and China on the other side of the old ideological curtain. A chorus of international denunciation of the South Asian rival followed, led by the U.S. India bore the brunt of America's "deeply disappointed" reaction to its defiance of the "international consensus about the need to promulgate a ban on such testing:

Was the American Central Intelligence Agency (CIA) aware of India's nuclear capacity and of its decision to demonstrate it to the world?

Apparently not, for the Agency came under intense criticism for its allegedly "colossal failure" to detect India's preparations for the explosions.

It was made clear that, had the U.S. known the tests were on their way, the Americans would have put intensive pressure on the Indian authorities to dissuade them from proceeding.

It could be argued, however, that India's explosions should not have taken the world by surprise, since Delhi had consistently refused to sign the Nuclear Non-Proliferation Treaty of 1967 or the more specific Comprehensive Test Ban Treaty of 1996. There was indeed a certain legitimacy about India's opposition to these international agreements on the ground that the outcome was, in effect, a kind of "apartheid" legitimizing the arsenals controlled by the existing nuclear powers while denying such weapons to powers that lacked them.

Before India's successful tests in 1998, India had claimed (in 1974) that such tests would be for peaceful purposes, whereas now the Indian prime minister's principal secretary, Brijesh Mishra, boasted that the three explosions provided his country "with a proven ca-

pacity for a weaponized nuclear program." The impression given to
the outside world was that the recently elected coalition government
under the Hindu nationalist Bharatiya Janata Party was determined
to follow a more aggressive foreign policy than its predecessors.
That this impression was indeed intentional had been made clear
after the election victory in March when the government announced
that it would change India's previously ambiguous nuclear policy.

At that time, Pakistan had immediately responded by testing its
medium-range surface-to-surface Ghauri missile, which was capable
of reaching deep into Indian territory.

Both rival powers made themselves deeply unpopular internationally, by their matching policies of competitive responses, internally, however, especially in India, the ventures into militant nuclear
testing made the successive governments sensationally more popular than they had been.

In a move clearly planned to reduce international concern over
the Indian-Pakistani mutual hostility, the Indian defense minister,
George Fernandes, declared on 3 May 1998 in a television interview, that the greatest potential security threat to India came not
from Pakistan but from China. In support of his argument, he accused China of providing Pakistan with missiles and nuclear technology, and with deploying nuclear weapons in Tibet along its border with India. On 5 May Beijing dismissed the Indian foreign
minister's charges as "absolutely ridiculous" and "not worthy of refu-
tation."

Faced with mounting international disapproval, Prime Minister
Vajpayee drafted a letter to six world leaders, plus Kofi Annan, secretary general of the United Nations, all dispatched on 11 May. The
major recipient was President Clinton of the United States. In it
Vajpayee gave factual support to his charges against China. He accused Pakistan of sponsoring a terrorist campaign against India in
Kashmir during the past ten years (in which, needless to say, he
ignored the provocation inherent in Nehru's insistence on incorporating the mainly Muslim country into independent India).

Next day, Clinton unambiguously condemned India for deliberately challenging "the firm international consensus to stop the pro-
liferation of weapons of mass destruction." Other critical reactions followed, from China and Kofi Annan. Japan—India's largest
foreign aid supporter—announced on 13 May that it would immediately freeze U.S.$25 million in grant aid to India. The day after, the

Japanese went further, suspending all further loans to India, totaling $988 million.

In a more roundabout move, President Clinton turned to the Nuclear Non-Proliferation Treaty, ratified by the U.S. in 1994, the terms of which required the U.S. to suspend all foreign aid to governments of nuclear-free states that indulged in nuclear tests. The other recipients of Vajpayee's collective letter were the leaders of Germany, Canada, Australia, Sweden, and Denmark, all of which announced that they would be cutting off economic aid to India.

Not unnaturally, international interest turned to Pakistan in the wake of the sensational news from India. In a telephone conversation with Clinton on 13 May Pakistan's prime minister, Nawaz Sharif, had declared that his country now had no option but to "take appropriate measures to protect its sovereignty and security." His foreign minister, Gohar Ayub Khan had been a trifle sterner, declaring that same day that India's actions "will not go unanswered."

Two days on, the U.S. president sent his deputy secretary of state, Strobe Talbott (long known for his tendency to placate the Soviet Union), on what was seen as a mission of dissuasion. With him was Anthony Zinni, commander of U.S. forces in the Middle East and Southwest Asia. While the option of sanctions was (naturally) discussed in Washington, before their departure, there was a realistic understanding in the White House that sanctions were not necessarily a desirable option. At the time, Pakistan's external debt totaled U.S.$ 32 billion, while its foreign reserves totaled only $1 billion. Indeed Pakistan had recently taken on an IMF loan totaling $1,558 million, for a three-year period.

Sharif set a conciliatory tone on 15 May when he declared that there was "no haste to test the bomb immediately." Two days later, however, Khan sounded a rather more hostile note, confirming that the Cabinet had already decided to respond to India's explosions with an underground nuclear test. The international atmosphere in the wake of India's tests (with the exception of the U.S.) had in fact been remarkably mild. France, Russia and the U.K. had all declined to consider what they described as a counter-productive package of sanctions. The final statement of the G-8 summit appealed to India to embrace the nuclear reduction treaties, and opted for a mere verbal reprimand to India's government.

On the 18th, India's Home Minister and deputy prime minister, Lal Krishna Advani (who would be placed in charge of the Department for Jammu and Kashmir Affairs the following week), warned Pakistan that it "should realize the change in the geostrategic situation" On 21 May however, in a speech to the Lok Sabha (the lower House) Vajpayee offered to negotiate a "no first use" nuclear pact with Pakistan.

The response came on 28 May when Pakistan announced that it had detonated five nuclear devices earlier that day at the Chaghai Hills testing site in the Baluchistan desert. Prime Minister Sahri thereupon asserted that, having done so, it had now "settled the score" with India.

It seems fair to say that with the Pakistani's comment, the India-Pakistan rivalry, often regarded by the outside world as extremely dangerous to its multiple interests, had reached a new low reminiscent of schoolboy rivalry. Regardless of Nehru's ancestors, India had lost Kashmir (or, more precisely, had failed to gain control over it). The India-Pakistan war of 1970-1972 had failed India in this respect, and Pakistan had lost its eastern territories, not to India, but to the Bangladeshis. There had been no political victory, for either side. The elusive prize had escaped hungry hands, yet again.

Part 3

Moscow's Afghan Obsession

10

The Tempting Country

Afghanistan could almost be said to have been designed by nature for conflicts, both national and international. A landlocked country, the nearest salt water is the Arabian Sea, 300 miles to the south. Apart from the fertile plains and foothills of the north, its south consists of sandy deserts and high plateaus, separated from the north by lofty mountains and valleys both deep and narrow. The mountains form an almost impenetrable barrier, with peaks of up to 21,000 feet.

Despite these forbidding characteristics, the country has been used historically as a path from one country to another and therefore, inevitably, as a temptation for foreign invaders. The military conquerors whose goal was India passed through Afghanistan, leaving in their wake what has been described, accurately if poetically, as "a mosaic of ethnic and linguistic groups"[1] The two competing major languages are Pushtu and Dari (a form of Persian). Migrations and conquests (the first being naturally the consequence of the other) have left linguistic heritages, including Uzbeks and Turks, but also on a smaller scale, Hindus and Sikhs, and a small minority of Jews.

As for religion, more than 90 percent of the 17 or 18 million inhabitants are Sunni Muslims; plus a few thousand Hindus or Sikhs. The ancient invaders of this complex country include the Greeks, the Iranians, the Turks and the Chinese, and Mongol hordes under Genghis Khan.

In modern times the British were among the aspiring conquerors, The British army entered Afghanistan from India in 1839, but was forced to pull out, mainly by guerrilla attacks. In November 1840, the British made a further attempt, but pulled out again in January 1842, after more and constant guerrilla attacks. The defining moment came on 6 January, when bands of guerrillas swarmed around 4,500 British and Indian troops, with some 12,000 camp followers, and massacred them.

Another year to remember was 1878: on 20 November, war broke out between Britain and Afghanistan, arising out of Britain's determination to secure its frontier with India in the face of Russia's advance into central Asia. With imperial authority, the British drove out the Amir, Sher Ali, and placed Yakub Khan on the vacant throne. In September 1879, the British agent Major Sir Pierre Cavagni, was murdered and there was a new flare-up. With Russian support, Abd ar-Rahman entered Afghanistan, but not very long after, on 20 July 1880, he reached an agreement with the British out of which he secured a peaceful old age, when the British rewarded him with a pension.

Murders in Kabul

In 1965, a partly pro-Soviet group was set up in Kabul. It called itself the People's Democratic Party of Afghanistan. Unlike so many parties using the magic word "People," it was neither totally nor unconditionally pro-Soviet, being in fact divided into two factions: a pro-Soviet one calling itself Parcham ("Banner") and a more nationalist one, Khalq ("Masses.")[1]

In July 1973, the Monarchy of King Zahir Shah had been toppled by his nationalist prime minister, Mohammed Daud. The fact that Daud had led a successful coup did not in itself bring prosperity or stability to the complex country. With its few natural resources, its nomadic population and pastoral society, a 70 percent illiteracy, and the resistance of tribes and sardars (chieftains) to any form of government, a successful coup did not necessarily bring authority to those involved.

Between 1973, the year of Daud's coup, and 1978 there were two attempts to overthrow him, both unsuccessful. Daud had doomed himself to being ousted, however, by launching a purge of local Communists. In 1977, his Minister of Planning was murdered in retaliation for Daud's arrest of Afghan Communists. In February 1978, apparently unaware of his self-inflicted peril, Daud arrested twenty-five Communists for conspiring against the government. Two months later, on 17 April, the leader of the pro-Moscow Parcham Party, Akhbar Khybar, was assassinated. His funeral was a signal for anti-government demonstrations, to which Daud responded by ordering widespread arrests of other Parcham figures. Some fled, others stayed but went underground.

Without the help of a military strongman, Daud might well have been ousted at this point. The strongman was Colonel Abdul Khadir, who had been involved in the 1973 coup that had ousted the King. Daud, desperate to stay in power, eased his strongman out. The Army had its own leader in mind and released him from jail. His name was Nur Mohammed Taraki, general secretary of the banned pro-Communist Khalq Party. It soon became clear that Moscow had been aware of the coup, although the Soviets did not necessarily order it. Its body, the Military Revolutionary Council (MRC) started purging the police and army of "reactionary elements."

At first, the MRC met with no visible opposition, but when it started acting against the Muslim League and provincial tribes, it ran into strong opposition from the League. Three months after the takeover, the MRC was divided down the middle between members of the Parcham and Khalq parties. When the Khalq was ordered by Moscow to work with the Daud regime, it had refused and at that point, the Parcham broke away to cooperate with Daud. The officer corps, however, had remained solidly pro-Moscow; not surprisingly, as its officers had all been trained in the Soviet Union.

On 27 April 1978, under pressure from Colonel Khadir, tanks and planes were deployed and the presidential Palace was stormed. Daud and his family were shot.

The Politburo instructed the KGB to produce a report on the Khalq and Parcham factions of the PDPA, and understandably, on the relative merits of Taraki and Karmal as "the most reasonable and disciplined of the two. In contrast, Taraki was seen as "stubborn, intolerant, irascible and shallow." According to the KGB defector Vladimir Kusichkin, the KGB's view was overruled by its ideological chief, M.A. Suslov, who had opted for Taraki.

One of Khadir's first actions was to release all three of the PDPA's top figures: Taraki himself, and his Khalq deputy, Hafizullah Amin, along with Babrak Karmal of the Parcham faction. As Suslov had advised, the Army appointed Taraki as president and prime minister. The outcome was a new ruling body, the Military Revolutionary Council, which named Khadir, by then promoted to brigadier-general, as defense minister, and Amin as deputy prime minister. The Constitution was abolished and the country was re-named the Democratic Republic of Afghanistan, in line with the constituent republics of the Soviet empire in Europe.

By the end of summer, a number of MRC members had been arrested for alleged "imperialist" plotting and "counter-revolutionary activities. The most prominent of them was Abdul Khadir himself, the former defense minister. After apparently admitting his failure to report an alleged counter-coup in September, he was charged with treason. Was he executed? The logical inference would be positive, although no record of his death emerged.

Other ministers were also arrested on similar charges. It was clear that all those arrested resented Soviet influence over Afghanistan, not least over the rapidly increasing number of Soviet advisers in the country. In fact, the number was reported to have risen from 1,500 to between 5,000 and 6,000 by early 1979.

On 5 December that year (1978), the Soviet Union and Afghanistan's pro-Soviet government signed a Treaty of Friendship: a political victory for Moscow, but not an enduring one.

As could have been predicted, there was growing hostility between the largely pro-Soviet Army and the unlinked and anti-Soviet tribes. A determinedly anti-Soviet National Liberation Front emerged in early 1979. (Of which, more later.)

11

The Fatal War (1980-1989)

Having done about all that could be done to bring Afghanistan under Moscow's political control, the Soviet Politburo had a difficult—in the end, fatal—decision to take. Should they use their overwhelming military force to bring that difficult country fully within the Soviet empire?

On 2 December 1979, the Soviet leadership had sent Lt-Gen. Viktor Paputin to Kabul, ostensibly to advise President Hafizullah Amin on internal security; but in reality mainly to persuade Amin to hand over power to their chosen protégé, Babrak Karmal, who had been sent to Prague as ambassador.[1] Another point was to make sure the Soviet Union was formally invited to send troops to Afghanistan, in line with the Soviet-Afghan Treaty of a year earlier, December 1978.

Not surprisingly, Amin was in no mood to take orders from Moscow. The conclusion to be drawn was clear: Amin would have to be removed, to make way for Moscow's obedient puppet, Babrak Karmal. How else could this be done other than by a full military takeover?

To deal with this and other problems, the Soviet Politburo met in Moscow on 12 December 1979, and reportedly decided to invade Afghanistan.[2] For months, there had been a build-up of Soviet forces along the Afghan border. Understandably, for ethnic reasons, most of them were Tadjiks or Uzbekis, and garbed in Afghan uniforms. That day, a Soviet military aircraft landed at Kabul airport, where a mixed commando unit launched an assault on the palace. The mixture was appropriate: a Spetsnaz unit from the GRU (military intelligence) and volunteers from the KGB's First Chief Directorate.

Their first mission was to seize political power at the centre. The assault units forced their way into the palace. After fierce fighting with Hafizullah Amin's highly trained special guard, they killed off the guard, trapped Amin in his room and killed him with a hail of bullets.

Not surprisingly, Moscow's choice to take over was Babrak Karmal, who had been moved to Prague as ambassador. Summoned to Moscow and appropriately briefed, he had broadcast a speech from the Soviet capital, on 27 December, declaring that he was the chosen president, that he would free political prisoners, respect the Islamic faith, and (in an unconvincing show of Western-style democracy) that he would allow a free press and free political parties.

Next day was to be invasion day. By then, Kabul radio was Soviet-controlled and announced that Karmal would be the overall leader: head of state, prime minister, chairman of the Revolutionary Council, secretary-general of the ruling party and commander-in-chief of the Armed Forces. On paper, at least, no definition of total power could be more complete.

Soviet forces crossed the border on 28 December 1979. Two days later, the Soviet Communist Party claimed through its official organ Pravda that these forces were "a limited contingent" to be "fully recalled" as soon as its mission was completed. What did these quoted words really mean? Western estimates turned out to be closer to the truth: by early January 1980, around 50,000 troops, with 1,000 tanks, had reached Kabul. When Western protests came in (notably from President Jimmy Carter of the U.S.) Moscow's reaction was that the Soviet Union was responding to an appeal from the Afghan government under the Friendship Treaty of 1978, in response to provocations from Afghanistan's enemies. Since an unfortunate speech by President Carter, Moscow felt it could reasonably disregard any risk of intervention from Washington.*

By the end of January, the number of Soviet troops in Afghanistan had reached 80,000. A few days earlier, on 25 April 1980 the new and anti-U.S. Islamic fundamentalist regime in Iran had seized sixty diplomats and other staff from the American embassy in Teheran. The U.S. had responded with a complex operation to free their compatriots, which ended disastrously when an American helicopter collided with a tanker aircraft in a sandstorm.

---

* "Being confident of our own future, we are now free of that inordinate fear of communism which once led us to embrace any dictator who joined us in that fear. We hope to persuade the Soviet Union that one country cannot impose its system of society upon another, either through direct military intervention or through the use of a client State's military forces, as was the case with Cuban intervention in Angola." (President Carter, in a speech at Notre Dame on 22 May 1977, quoted in +War Called Peace, p.227).

Realistically, the Soviet leaders felt that their invaders of Afghanistan could proceed with their military and political plans without fear of an American response. Understandably the Soviet invasion force was largely drawn from Soviet Turkestan and Central Asia, enabling them (at least in Soviet military planning eyes) to merge with the local northern Afghan population in racial appearance and Islamic custom.

As always, the Soviets had been using their long-established technique, known as "disinformation." The Afghan government had obviously been led to assume that the newly arrived Soviet forces were there to strengthen their own authority. The Soviet disinformation machine had led them to believe that Afghan army units were about to be re-equipped by the Soviet army. They were persuaded to disarm their tanks in the belief that new equipment was on its way from Russia.[4]

Not all the Afghan troops gave in so easily, however. In some areas, irrespective of reported political arrangements between their government and the Soviet government, the Afghan Army resisted. Among the exceptions was the 8th Afghan Division, which fought hard against the invaders, losing 2,000 men. But resistance was the exception to the rule. Perhaps as many as half the Afghan aarmy's 80,000 men were reported to have deserted. Most of them went home to their villages, but some joined their compatriots' freedom fighters.

Although the leftists put in power by the Soviets cooperated with the invaders, guerrilla groups in many areas fought a war of ambushes followed by disappearance into the mountain wilds. Soviet soldiers unwise enough to wander alone in the villages would be killed by stones and knives.[5]

Effective though guerrilla methods were, it was virtually impossible to merge the resisters into organized army units. They lacked modern weapons and were unable to coordinate their isolated attacks. On the other side, the Soviet occupiers found it almost impossible to coordinate their defensive measures.

Thus the war dragged on, year after year, while the various mujaheddin groups failed to organize a centrally controlled resistance army. Early in July 1980 the Soviets had stepped up their deliveries of suitable equipment, Hundreds of Antonov transports landed at Kabul and the military airport at Bagram. With them came light tanks, scout cars and armored vehicles, and some helicopter gunships. For infantry use, many AK-S assault rifles (an improved version of the AK-47, and specifically designed "nasty" bullets, including the notorious dumdum bullets banned under international law, and cluster bombs scattering needle-sharp arrows over a wide area.

Contrary to Moscow's initial optimism, the war in Afghanistan dragged on and on. In late December 1986 (as Alex E. Alexiev, the Rand Corporation's senior analyst on Soviet affairs commented,6 the war in Afghanistan entered its eighth year. As Alexiev observed: "...This not only brings it close to being twice as long as the 'Great Patriotic War', as the Soviets call World War II, but, also makes it the longest counter-insurgency war in Soviet history."

As he noted, the Basmachi rebellion of 1919-1926, in Central Asia was defeated after more than seven years of bitter fighting. This was also the outcome of resistance wars in the Western Ukraine (1944-1950) and Baltic (1945-1952).

Clearly, there was no danger that Afghan resistance would give in to the Soviet occupiers of their country, no matter how long the clash went on. To define Soviet strategy, as it was seen in Moscow, is worth attempting. As Alexiev states, Western analysts tend to confuse Soviet strategy with military objectives and behavior. Undoubtedly, for observers of Soviet methods, political and economic objectives played a more important role in Soviet military thinking than it did in Western strategy (as indeed is confirmed by other analyses in this book).

Initially, Soviet military as well as political leaders may well have believed that they could achieve a relatively quick solution. After the conflict continued, well beyond the year, it became clear to Moscow that military strategy alone would not bring victory.

Typically, in Communist methodology, the aim was not just military victory (even if this were to prove possible) but political solutions in line with Communist theory. Clearly, the urban population had to be given priority over the peasantry, as more likely to fall in with Moscow's ideas. However, in view of the potential long-term challenge from the peasantry, it would be necessary to make any establishment of liberated territories impossible. Clearly, the civilian population would have to be terrorized to stop its support to the resistance; or failing that, to drive as many civilians as possible into leaving the country, taking refuge abroad. And, not least, cripple the mujahideen by cutting off its supply routes.

To this end, the Soviet invaders had been establishing widely mined secure perimeters, alongside major highways and especially around the cities. Indeed such methods had yielded apparently encouraging results, in that the Soviets had secured not only Kabul, the capital, but also the next two largest cities: Jalalabad and Mazar es-Sharif. However, against that, they had failed, by the winter of 1987, to bring two other major provincial centers under their control: Kandahar and Herat. In both these areas, the resistance had consolidated its position.

Not all Soviet efforts had failed, however. Following Mikhail Gorbachev's advent to power. in 1985, the Soviet military effort increased spectacularly, to some extent by copying and adapting guerrilla methods. To quote Alexiev again:

"Utilising commando-type units, such as airborne, special assault, and independent reconnaissance forces, the Soviets have become increasingly adept at operations behind enemy lines, night ambushes and surprise attacks."

With their usual indifference to Western reactions in a war situation. the Soviets continued their policy of deliberate destruction of villages, indiscriminate carpet bombing, execution of civilian hostages and booby-trapped "toy bombs." When captured, hostile mujahideen were normally summarily executed. To quote Alexiev yet again:" Contrary to the prevalent journalistic cliché, Afghanistan is not a Soviet 'Vietnam'. Whereas American expenditure on the Vietnam war, at its height, was between 25 and 30 percent of total U.S. defence expenditure, the comparable Soviet expenditure in Afghanistan was less than 3 percent."

With their usual skill in such matters, the Soviet war planners concentrated on two distinct objectives. One was to transform this war of invasion into a civil war; the other was to provide the existing regime in Kabul with the elements of legitimacy in the eyes of the outside world.

As always, the guiding principle, for the KGB, was to cover up the reality of the conquerors' hold on power by training and indoctrinating Afghan cadres, both civil and military, to run what parts of the country they controlled in a convincingly effective way. However efficient and convincing the cadres were, they would, of course, continue to be controlled by the Soviet invaders. By late 1985, more than 10,000 Afghan military and intelligence officials had returned to Afghanistan after thorough training in the Soviet Union and Eastern Europe. Training and teaching facilities abroad were given priority over those available in Afghanistan itself. For ten years at least, children aged ten and upwards were sent to Soviet Union schools. By the mid-eighties, there were more Afghan students at Soviet universities than in their own country's.

As in the east European satellites, the Afghan Khad (equivalent of the KGB) had reached a degree of surveillance efficiency nearly equal to that of their KGB masters. By the mid-eighties, the Khad membership had reached 30,000.

At a time when the Western world was falling under the remarkable personal spell of the new Soviet leader, Mikhail Gorbachev, his own policies of "broadening the democratic base of Afghan revolution" had cast its spell over Afghanistan's more modest center of power. To carry out (or give credence to) this democratic base, the ruling party was required to make conciliatory gestures towards Islamic leaders, businessmen and professionals, while consciously toning down the inevitable Marxist rhetoric characteristic of all Communist parties in power. Despite the passive adoption of Soviet precepts, it would be mistaken to suppose that this policy of obedience really consolidated the Communist victory. In fact, it increased Islamic and tribal resistance to the new regime.

12

Soviet Methods

Typically, the Soviet apparatus used its known and effective skills within Afghanistan and abroad, to disguise the nature of the war it was waging. It has to be added, however, that these skills were at their most effective in America during the Carter administration.

As Alex Alexiev pointed out,1 their methods can be listed as follows:

Dividing the tribes. They co-opted support from certain tribal chiefs, while antagonizing others. They did so by offering money and weapons to a selected tribe, in return for support against rival tribes. Alexiev gives a significant example: closing off a key supply route to the northern area of Afghanistan by persuading the tribe to set up a militia, while denying the anti-Soviet mujahideen passage through the Nuristan Valley territory in the winter of 1987. In one case, they co-opted tribal forces on the Pakistani side of the border, thus endangering mujaheddin infiltration points.

Removing the Afghanistan issue from international public opinion, while persuading Pakistan, by a mix of threats and promises, to withhold support for the resistance while denying refuge on Pakistani territory: a vital element in tribal resistance. As always, the vast Soviet propaganda apparatus was used to persuade the West that the ultimate Soviet aim was a negotiated settlement, while letting the West understand that the vitally important arms control agreements would be undermined by raising the Afghanistan issue.

As always, encouraging noises to the tribes were balanced by ruthless measures, such as the destruction of crops by napalm and anti-personnel mines, destruction of the irrigation system, buying food from the peasantry at higher than market price to deprive the mujahideen of their own expected supplies, while providing peasants in government-controlled areas with seed and fertilizer.

And again, as always, a large part of the most productive land was rendered useless by heavy mining of selected areas. In various places hostile mujahideen were left no choice but to import food from Pakistan at exorbitant prices.

Nevertheless, in the face of such ruthless measures, the mujahideen expanded rather than contracted in defiance of the foreign enemy, and consistently with the ever increasing professionalism of the mujaheddin, as the war years lengthened. Interestingly, in the years 1985 to 1987, the mujahideen were able to seek and obtain long-range weapons, such as the Chinese-made 107mm multiple rocket launcher.

In the long run, no doubt, the best and most durable asset on the mujahideen side was their indomitable courage. As Alexiev pointed out, the biggest military problem faced by the Afghan resistance was the almost total lack of modern anti-aircraft weapons. Many tribesmen disposed of 12mm machine-guns, Soviet-designed but useless against fatal attacks by armored Soviet helicopter gunships. Guerrillas armed with SAM-7 weapons were fortunate, but apparently few.

One long-range weapon in use by the mujahideen was the Russian 82- mm mortar which, in its day, was considered impressive; but that "day" was during the latter phase of World War II: its range was only three kilometers, and it was not noted for its accuracy.

By turning official eyes away from the Afghan scene, the Carter government played into the hands of the Soviet invaders. Meanwhile, after seven years of a superpower invasion, the Afghan people had suffered about a million killed and five million driven from their peasant homes.

This sad picture of events, though published in 1988, largely reflected the situation that had developed during the Carter term in office. He was, of course, defeated by Ronald Reagan in the 1980 presidential election, but the advent of the determinedly anti-communist president did not produce immediate results on the Afghan scene, for reasons explained by the same experienced observer, Alexiev. Shortly after the Soviet invasion, a senior Carter administration official expressed himself in these words: "The question was, do we give them [the insurgents] weapons to kill themselves, because that is what we would be doing. There is no way they could beat the Soviets."[2]

The author quotes other comments from government officials, as late as 1985, to draw attention to the marked reluctance of U.S. officialdom to encourage the United States to provide effective help to the anti-Soviet guerrillas. However, by the second half of 1984, support for the Afghan anti-communist cause had become "virtually unanimous and truly bi-partisan" in the U.S. Congress, especially the Senate, Despite determined opposition from the State Department and the Central Intelligence Agency, both Houses of Congress passed a strongly worded resolution in November 1984 calling for the supply of effective weapons to the guerrillas.[3]

In the fiscal year (1 October 1984 to 30 September 1985) aid to the mujahideen by the U.S. Central Intelligence Agency (CIA) had risen to $250 million: more than 80 percent of the Agency's annual budget for covert operations. A further $250 million in covert military aid was alleged to have been secretly approved by the Congress in September 1985.

The following April, President Reagan signed a national security directive, NSDD 166, stipulating that it was henceforth U.S. policy to help the mujahideen drive out the Soviet forces "by all available means."

By 1987, there had been a dramatic increase in U.S. assistance, totaling $680 million. The turning point was marked by the first deliveries of sophisticated anti-air missiles.

A year later, in the Winter of 1988, the same realistic observer of the Soviet war on the Afghan people, Alex Alexiev, was able to draw encouraging conclusions from the radically changed attitude ot the Reagan government towards Afghanistan. In the autumn of 1986, substantial deliveries of the U.S.-made Stinger missiles and of the less effective British Blowpipes between them turned the tide in favor of the mujahideen. Not long after, kill ratios of nearly 50 percent turned the tables on the Soviet aggressors.

Within a year, 270 Soviet aircraft had been destroyed, at a cost to the Soviet Union of around $2.2 billion.4 Not least, a corresponding number of Soviet pilots would have lost their lives or capacity for service.

As it happened, the policy change of April 1985 in America coincided, almost exactly, with Mikhail Gorbachev's access to power as general secretary of the Soviet Politburo. Despite the new Soviet leader's propagandist insistence on his favorite catchwords (such as "glasnost" or openness, and "New political thinking") his advent to power was followed by an immediate intensification of the war in Afghanistan. Contrary to his presumably optimistic calculations, however, there was no sign that the war would end in a Soviet victory. Instead, the improved resistance of the Afghan guerrillas gradually reduced any military optimism on the Soviet side.

At a meeting of the Soviet Politburo on 20 March 1986, Gorbachev, who chaired the meeting, read out a pessimistic memorandum on the situation, recording that the then Afghan pro-Soviet leader, Babrak Karmal, was by then "very much down in terms of health and of psychological disposition." After a further trip to Moscow on 31 March, Karmal wrote to the Soviet Politburo on 4 May, asking to be relieved of his duties on grounds of "ill health."

Following a meeting of the Politburo on 13 November 1986, Gorbachev complained to his colleagues that the goal set in October 1985—to expedite the withdrawal of the Soviet forces while ensuring a friendly Afghanistan—had not been met. He went on to call, not for the first time, for the withdrawal of 50 percent of the Soviet troops from Afghanistan in 1987 and of the remaining 50 percent the following year. Although not in so many words, this pessimistic statement amounted to an admission of defeat.

Not all Western observers would have agreed with this form of words at that time. In my column in the New York-based *National Review* of 27 October 1988, I wrote: "In strategic and historical terms, Mikhail Gorbachev's decision to pull the Soviet forces out of their Afghan quagmire constitutes a major reverse for the October Revolution, comparable to the American decision to withdraw from Vietnam in 1973. In each case, the superpower involved tacitly conceded defeat."

At that time, not every observer would have agreed with this verdict. But in fact, the Soviet Union itself crashed a mere two years later.

13

The Rise of the Taliban (1995-1996)

The departure of the Soviet forces in 1989, defeated though virtually intact, left behind an unparalleled state of disorder, confusion, anarchy. For years, hostile rivalries between competing Islamic tribes were marked by continued fighting and brutalities, each and all in the name of religion and nationalist claims.

On 21 November 1990, President Najibullah called a press conference at the United Nations offices in Geneva, at which he confirmed reports that he had held talks with "prominent personalities of the opposition side" in search of a political solution to end twelve years of civil war. He was careful to praise the "new, positive and changing" attitude of the United States. Presumably, it was thought at that time, this was a muted praise for recent U.S. Congressional moves to cut aid to the mujaheddin by 10 percent.

He had been vague about the "prominent personalities" he had met. That day, one of the tribal leaders, Seghbatullah Mujjadeddi, president of the Pakistan-based Afghan Interim Government, denied either meeting Najibullah or aiming to meet him, in discouraging words: "Neither we nor our representatives are ready to hold talks with the puppet Najib or his representatives."

Finally, after 14 years of fighting, the Mujaheddin guerrillas toppled the Soviet-backed Najibullah, on 15 April 1992. He had attempted to flee the country as the hostile guerrillas closed in on him, but was halted on the way to the airport by troops under the command of the Uzbek military leader, General Abdul Rashid Dostam. In earlier days, the Najib government had used Dostam's force against the Pashtun mujaheddin Fierce fighting followed on 26 April, involving tanks, artillery and aircraft. Three days later, an interim government of the new Islamic State of Afghanistan was formed under the Presidency of Mujjaddedi.

One of the first acts of the new government was to announce an amnesty for all members of the old government, with one exception: Najibullah.

The formation of a post-Soviet government may have seemed promising at the time, but internal skirmishes between the rival tribes continued on a mounting scale, for years after the departure of the Soviet Army. And then, early in 1995, a new name came into the news: the Taliban, apparently an army of students, devoted to total Islamic orthodoxy and determined to achieve complete power. By mid-February the Taliban had brought 10 of the country's 30 provinces under its control and was now within fifteen kilometers of Kabul.

On its way to the capital and to political power, the Taliban had dismantled mujaheddin roadblocks. Not only had it disarmed whatever gunmen stood in its way, it had also ruthlessly suppressed the opium and heroin trade. On 13 February its members had driven rival mujahideen out of the main convoy route from Pakistan to Kabul.

New headlines were on their way, notably when the Taliban seized the headquarters at Charasiab of the most formidable tribal leader until then, ex-Prime Minister Gulbuddin Hekmatyar, who headed the Hezb-i-Islami. Already, on 10 February, the Taliban had taken Hekmatyar's stronghold at Maydan Shahr, capital of Wardag province, 30 kilometers west of Kabul; and three days later they had likewise seized another Hekmatyar stronghold at Pul-i-Aslam, capital of Loghar province.

Meanwhile, on 13 February, government forces had recaptured the strategic northern town of Kunduz from the Uzbek army of General Abdul Rashid Dostam, Hekmatyar's ally, which had taken possession of it on 5 February.

One thing the Taliban forces had in common with Hekmatyar's was the fact that they were mainly ethnic Pashtuns from the Durrani and Ghilzay tribes of southern Afghanistan. In tribal contrast, the army loyal to President Burhanuddin Rabbani was predominantly Tajik.

Every day brought more background about the Taliban, who had emerged from obscurity to dominate the scene. Predominantly, it was reported, they were Sunni Muslims and before forming their military organization in late October 1994 most of them had been students at Muslim religious seminaries, known as hadrassas, in northwestern Pakistan. They were reported to have built up links with foreign governments, the names of which were kept secret, but whose contribution to their forces was serious, since they were armed with sophisticated weaponry including ten fighter aircraft and 200 tanks.

Until then, nobody had heard of the Taliban leader, whose name and background were revealed on 17 February 1995: Muhammad Umar, from the southern province of Urozgan. His followers made it clear that he was dedicated to the elimination of all other armed factions and the creation of a strict Islamic state, excluding all women from public life.

At this time, the UN peace plan had been in the news: it involved the transfer of power to an interim council. But suddenly, on 22 February, it was pushed into the background when it was announced that Rabbani had decided to postpone the transfer until 22 March. In fact, Rabbani had let it be known that he would resign on the 21st, and indeed had pledged that he would not impose any further pre-conditions, so that fresh discussions could begin on the formation of a new interim council.

The interesting point, however, was that (according to other reports) Rabbani had insisted on staying in power until the Taliban were included in the interim council. To that end, a Taliban delegation had met a representative of Rabbani in Kabul on 18 February. There, the Taliban representatives had spelt out three conditions for participating in the peace process: the restriction of interim council membership to "good Muslims," the representation of all thirty of Afghanistan's provinces and the policing of Kabul by "a neutral force" under control of the far-from-neutral Taliban.

On the 21st, a government spokesman in Islamabad (Pakistan) said that the UN was against the involvement of the Taliban in the interim council, on the ground that they were "a different sort of force from the Afghan parties." He went on to allege that the importance of the Taliban had been exaggerated by rival Afghan factions intent on delaying the peace process.

Clearly, this was not the kind of statement that the Taliban wished to hear. Not surprisingly, then, peace talks between government representatives and the Taliban, held near Charasiab (outside Kabul) ended on 28 February without a settlement on the disarming of rival factions.

After the newsworthy February successes, the Taliban may well have thought that power was "round the corner," but a fortnight on they faced an unexpected reverse at the hands of the Rabbani government. Heavy fighting had broken out in the Karte Seh district of south-west Kabul on 6 March. The Taliban were not involved in it, but suffered as a result.

On 6 March, the Army had attacked positions held by the Shia Hezb-i-Wahdat-i-Islami faction. The government used fighter-bombers to strafe and rocket Hezb-i-Wahdat positions, with the apparent aim of removing the last remaining opposition group from Kabul. In effect, the Hezb-i-Wahdat fighters were trapped in Karte Seg because the Taliban, by now in control of the capital's outskirts, had cut off the faction's supply lines and escape route,

Defeat was looming for the Hezb-i-Wahdat, whose leader, Abdul Ali Mazari struck a deal with the Taliban leaders on 8 March to allow Taliban men to occupy his faction's front-line positions. Not unexpectedly, the Taliban immediately took advantage of its new location to attack the Army. Already, the initial fighting had left around 100 dead and 1,000 injured, most of them civilians. On 12 March, the Army launched an offensive and pushed the Taliban force out of Karte Seh and back towards its Caharasiab headquarters. This was the first major Taliban defeat. For no strategic reason, the retreating Taliban force launched rocket attacks on Kabul, killing several civilians but without any military advantage.

As always in these tribal wars, one attack followed another. On the 13th, the Hezb-i-Wahdat leader, Mazari, whom the Taliban had taken with them as they retreated, was killed. The Taliban claimed that he had launched a helicopter attack on his Taliban escorts. On the 19th, Iranian radio reported that Hezb-i-Wahdat had appointed a "provisional leader" of their faction: Karim Khalili.

Four days earlier, on the 15th, Taliban forces had captured the Qala-i -Hyder base at the western entrance to Kabul. A short-lived success, for the Army took it back after a ferocious eight-hour battle.

On the 19th, the Army ousted the Taliban from its headquarters at Charasiab into Logar province: a retreat with one advantage: it was well out of rocket range from Kabul. The Taliban was not in a resting mood, however, and two days later it launched a further counter-offensive, halting the Army's advance at Mohammad Aggha.

Three days earlier, there had been another setback to the UN peace process, when Rabbani announced that although he still supported the process, he was "waiting for a reliable mechanism" before transferring power. On 20 March, the UN peace envoy, Mahmoud Mestiri declared that the transfer was postponed for a further fifteen days.

Although heavy fighting continued between the Taliban militia and government forces, a temporary peace had returned to Kabul, and on 3 April the Foreign Ministry invited various countries to re-open their diplomatic missions in the capital. Several countries responded to the invitation. One was India, on 3 May. Another, next day, was Pakistan. The U.S. and U.K. missions did not respond immediately, and on 4 and 22 May respectively, officials in Kabul met with members of these missions based in Pakistan.

After further fighting, the government of President Burhanuddin Rabbani signed a ten-day truce with the Taliban on 9 June 1995, and on the 29th Iranian radio reported that new agreements had been reached, covering the extension of the cease-fire and the release of all prisoners of war. The main reason why so much precise news on Afghanistan was coming from Iran was that, at that stage, some 1,600,000 Afghanis had taken refuge there. Following a tripartite agreement between Afghanistan, Iran and the Office of the UN Commissioner for Refugees resulting in an initial repatriation of 400,000 Afghan refugees. Preparations were also under way for the repatriation of a further 500,000.

July brought no fewer than four separate peace initiatives, sponsored by the UN, the Organization of the Islamic Conference (OIC), Pakistan and Saudi Arabia None of them was successful. The UN peace envoy, Mahmoud Messiri, had talks in Kabul with :President Rabbani, who urged the adoption of a new government peace plan involving an Islamic *loya jirga* (council) comprised of representatives of the universities, the provinces, the High Court and the Academy of Sciences. A logical plan, but too complex for rapid acceptance.

The OIC had first launched its initiative in July 1994. Now, a year later, the OIC sent a committee of four to Kabul, with the same proposals for a peace conference in Jeddah, Saudi Arabia. Once again the Afghans objected, as before on the ground that "peace activities should be concentrated in Afghanistan."

An earlier Saudi Arabian proposal, presented on 9 July, was much the same as OIC's: peace sessions in Jeddah under a joint supervisory committee, organized jointly by Saudi Arabia and Pakistan.

As for the Pakistani initiative, it merely aroused official anger over a visit to Pakistan in early July by General Abdul Wali, whose impediment was that he was the son-in-law of the ex-King Zahir Shah, in exile in Rome. Wali had reportedly lobbied for a role for the ex-monarch. He too had urged the revival of the traditional *loya jirga*, which as before was turned down instantly by mujaheddin groups opposed to any participation of the tribal and landowning elites.

So, talking had got nowhere; so fighting resumed.

August 1995 brought the Taliban an unpredictable spot in the news, when a group of its fighters hijacked a Russian Ilyushin-75 cargo aircraft in Kandahar on 3 August. The plane, on a flight from Albania to Kabul, had apparently been chartered by the Afghan government to bring arms purchased in Albania. The Taliban hijackers claimed that the arms had been bought illegally and were being transported secretly. Soviet officials strenuously denied the allegation. On the 27th the hijackers called for a Russian undertaking not to interfere in Afghanistan's internal affairs. How the dispute ended was not made clear.

Towards the end of August, the Taliban repulsed a government seizure of Geresh, recaptured the town and advanced towards Herat, 25 kilometers northwest. On 2 September, Taliban forces took Farah and next day they overran Shindan air force base, the largest in the country, seizing government aircraft. Herat city fell on the 5th. To the pleasant surprise of the Taliban, they had met no resistance from the forces of the provincial Governor, Ismail Khan, an ally of Rabbani.

For the Taliban, the seizure of Herat was regarded as of special interest, for it was the first Farsi-speaking city to fall into the hands of the predominantly Pashtun Taliban. Neighboring Iran was alarmed at the news of the capture of Herat, in fear of unrest close to its border.

These latest Taliban successes were attributed in part to a decision by the Uzbek leader, Abdul Rashid Dostam, to order his air force to bomb the city and cut the road link between Herat and Turkmenistan, after gaining control of the north-western province of Badghis. A complicating factor in the maze of rivalries was a recent alliance between Dostam and the independent-minded prime minister, Hekmatyar.

To complicate the situation still further, rumors circulated alleging Pakistani support for the Taliban. The rumors were denied, but not convincingly enough to stop a 5,000-strong mob attack on Pakistan's embassy in Kabul, killing an employee and injuring the ambassador and twenty-five others. The attack followed an official Afghan protest to the UN, accusing Pakistan of interfering in Afghanistan's internal affairs, and assisting "its puppets, the Taliban, to dominate parts of western Afghanistan."

On 21 September, Pakistan retaliated by ordering the expulsion of thirteen Afghan diplomats, without stating its reasons for so doing. On 10 September, Pakistan's prime minister, Benazir Bhutto, had sought to restore good relations by declaring that Pakistan would not break diplomatic relations with Afghanistan, but stressing that the embassy would remain closed until firm guarantees of security were given. Iran, however, closed its border with Afghanistan on 6 September and halted the earlier agreement on the repatriation of Afghan refugees.

In mid-September, the Taliban launched a series of rocket attacks against government positions in Kabul, from their bases in the hills surrounding Maidam Shahr, forty kilometers west of Kabul. The attacks went on, notably culminating in a rocket strike on a crowded marketplace in Kabul, killing seventeen people and injuring another twenty-six.

Next month saw a sharp escalation in the civil war, when Taliban fighters, with air attacks in support, launched a massive attempt to seize control of Kabul. The Taliban had seized Charasyab, twenty-five kilometers to the south, as a base for the assault. At least 550 people had been wounded, but there were no available figures for the dead. In late November, the Taliban intensified the crisis by destructive air attacks on residential districts of Kabul, killing nearly forty people and injuring a further 140.

They intensified their assault on the capital with more rocket attacks. A serious reverse occurred on 14 November when the one-eyed Taliban leader, Muhammad Umar, died in a helicopter that exploded in mid-air over Wardag province. Nor was he the only leader to die. Another was Mullah Abdul Salaam, notorious for the kidnapping of Pakistani and Chinese hostages in June 1993, was killed in clashes in the southeastern province of Zabol that same day: 14 November.

A week earlier, President Rabbani had offered to resign, but only on condition of an immediate cease-fire by the Taliban and an end to "foreign interference": words clearly aimed at Pakistan.

In mid-December Taliban forces escalated their attacks still further, after being forced to retreat some twenty kilometers south of Kabul. On the 12th, they shelled the city centre, and on the 19th, they fired nearly sixty rockets into Kabul's residential areas, with further attacks on 25, 27, and 30 December.

On 7 March 1996, the Hezb-i-Islami party, still led by the former Prime Minister Gulbuddin Hekmatyar, signed a military agreement with President Rabbani for joint action against the Taliban. One side-effect, widely welcomed in Kabul, was the restoration of electricity in parts of the city, after a two-year break arising from the control seized by Hekmatyar's forces over a large hydro-electric plant at Manipur, thirty kilometers east of the capital.

Hekmatyar's limited deal had no effect on the fierce fighting around Kabul during March. On the 4th and 5th, the Taliban launched a rocket attack on residential areas of Kabul, killing half a dozen people. The government retaliated with bombing raids on Taliban positions, on 13 March. The result: sixteen Taliban militia deaths. Heavy weapons fire exchanges continued. On 24 March, eighteen civilian shoppers were killed, and fourteen wounded in a busy Kabul street.

A Defense spokesman blamed the Taliban for the attack and the usual retaliation followed two weeks later when the Afghan air force bombed rebel positions around Kabul, killing up to 50 people. Twenty more were reported to have been killed when government planes attacked Charasyab where Taliban commanders were meeting.

The postwar civil war went on, relentlessly.

On 16 April, an unprecedented verdict issued by the Afghan Supreme Court broadened, at least temporarily, popular talking about the headlines. That day the Court ruled that "the appointment of an amir al momineen" (leader of the believers) by the "so-called Taliban" was un-Islamic on the ground that he had only one eye and that "the appointment of a person whose vision is defective is not allowed under Islamic law."

This was a curious tit-for-tat response to an interview in the London-based Arabic newspaper Al Hayat, in which the Taliban leader, Muhammad Umar, thirty-eight, acknowledged the title of amir and revealed that he had lost one eye while fighting the Soviet invaders in the earlier war. He had given the interview in Kandahar, partly to disprove reports that he had been killed in a helicopter crash in November.

Any observer of the military scene at that time might well have dismissed any suggestions that the Taliban might win the battle for ruling power in Kabul, for on 13 April, Rabbani's loyal troops had begun a major offensive to recapture the Western city of Herat and were said to have wrested control of most of the Western province of Ghor from the Taliban.

On 24 May 1996, President Rabbani signed a peace agreement with the Hezb-i-Islami leader, Hekmatyar. The two sides agreed to work together to organize fresh elections and establish a "real Islamic government." This was clearly a challenge to the extreme Islamic claims of the Taliban. It came with the entry of hundreds of Hezb-i-Islami forces into Kabul.

A few weeks later, at a ceremony in Kabul on 26 June, Hekmatyar was formally reappointed prime minister by President Rabbani: an interesting peace of reconciliation between the two former political foes.

The Taliban, clearly enraged by Hekmatyar's return, launched a ferocious rocket attack against Kabul, killing more than fifty people and wounding at least 130. (Not, it would seem, an attempt at endearing itself to the people the Taliban aspired to rule.) On 3 July, President Rabbani named a new government, headed by Hekmatyar as prime minister. His Hezb-i-Islami was awarded the key ministries of Defense and Finance, but Rabbani retained an important slice of formal power, as his Jamiat-i-Islami retained Foreign Affairs and was given the Interior Ministry.

The military, political and constitutional crisis came in September 1996, when the Taliban gained control over Kabul Two swift actions followed, and one major, protracted one. The first of the swift actions was the ousting of President Burhanuddin Rabbani and his government. The second was the summary execution (without trial) of the former Communist president Najibullah, which took place on 27 September. First, he was abducted from the UN compound in Kabul. Without ceremony, he was then shot dead, and his body was hung from a traffic kiosk near the presidential palace over which he had presided. For good measure, the Taliban forces also summarily executed Najibullah's brother and two of his close associates who had also found a very temporary refuge in the UN compound.

The UN special envoy to Afghanistan, Norbert Holl, strongly condemned these killings, expressing "deep dismay" that they had been carried out without a trial. The interim government neither apologized nor expressed any self-criticism, but simply "justified" the execution of the former president as "an Islamic act."

The major protracted change inflicted on Afghanistan was the transformation of the whole country, by stages, into a totalist Islamic State, run by men only, with women turned into virtual slaves, without the right to show their faces in public.

Some months earlier the Communist regime, never fully consolidated, had collapsed, in accelerating stages. On 18 March, Najibullah had formally announced that he was ready to resign, ostensibly to further the progress of the UN peace plan. At that time, the UN special representative for Afghanistan, Benon Sevan, had expounded that plan, the essence of which was to establish an interim, all-party government. As so often with UN schemes for reconciling mutually hostile factions, even when the hostility seemed irremediable, the situation did not encourage the slightest optimism. Certainly there was no consensus among the mujaheddin tribes for the formation of an interim regime or on the elements of a peace plan.

The situation was as follows:

1. In the north, the mujaheddin contingent had just captured the strategic city of Mazar-i-Sharif and other strongholds. The leader in this area was Ahmed Shah Masud, a Tajik commander fighting under the banner of Rabbani's moderate Jamiat-i-Islami. Masud had successfully arranged a series of local alliances with pro-government militia leaders from the non-Pashtun minority groups. Masud's forces were closing in on Kabul, in the face of little opposition.

Meanwhile, rival fighters from Hekmatyar's hard-line Hezb-i-Islami . were also closing in on the capital, from the south and east By early April, both forces were ready to take Kabul, while the UN's Sevan was desperately trying to broker a peaceful transfer of power.

From his "observation post," Najibullah had been increasingly desperate to flee the country. He headed towards the airport but was stopped by troops under the command of Gen. Dostam, the Uzbek commander mentioned earlier, who, by this time had switched allegiance to Masud. Failing to escape, Najibullah took refuge in Sevan's UN office in Kabul.

When Soviet forces completed their withdrawal from Afghanistan in February 1989, Najibullah's downfall was widely predicted —indeed, taken for granted. But, contrary to predictions, he remained in office, even surviving a serious coup attempt in 1990.

The Taliban's final advance on Kabul had come after a series of striking military gains in early September. On the 11th, they had overrun the eastern province of Nahgarhar and seized its capital, Jalalabad, 110 kilometers east of Kabul. Because of the town's strategic importance as the main conduit for supplies to Kabul, its fall had been a major blow to the government. The outgoing cabinet accused Pakistan of siding with the Taliban and "hatching a conspiracy to wage war on Jalalabad." Not surprisingly, Pakistan strongly denied these allegations.

The restrictive measures affecting women need to be recalled. New laws were passed, compelling women to stay at home, and banning them from seeking jobs or studying in schools. All women appearing in public were forced to cover themselves in full-length *burqas* (black veils). The few women who sought to assert their former rights to appear unveiled, or in some other way to be inadequately clad were beaten in the street. Any woman who committed thefts had a limb amputated, without anesthetic, Those accused of committing adultery were stoned to death. As for men, whatever their taste for whiskers, they were compelled to grow beards. Moreover, there was an immediate ban on films, television, cassette recorders and video players, all deemed to be irreligious.

Two days after the Taliban achieved full power, its regime was recognized by Pakistan (on 29 September). The United States expressed support for the new regime, though stopping short of full recognition. The Soviet Union merely emphasized that it would not intervene in Afghanistan. Its one, failed intervention was deemed to be enough.

In early October, the new regime, which had barely come to power, came under attack from a strong alliance of opposition forces controlled by Ahmed Shah Massud, the former defense minister and government military commander. Pakistan was ready to mediate, as the only foreign power to have recognized the new regime, but its attempt was brushed aside. The Taliban forces advanced towards the Panjsher Valley and the strategic Salang Pass, but were forced to retreat on meeting fierce resistance from the combined forces loyal to Masud and the northern Uzbek leader, General Dostam. On 15 October, a government spokesman admitted that the Taliban forces had lost control of key areas north of Kabul to Masud's forces, but claimed that the retreat was "temporary."

Once again, the Taliban's claim was proved well founded, when they recaptured the town of Qarabagh, fifty kilometers north of Kabul, on 16 October. They had also recaptured Qala Nou, capital of the Badghis province in the West, on the 17th. Once again, however, their victories were short-lived, for on the 20th, Masud's forces were reported to have recaptured Qarabagh and the strategic air base at Baghram. By the end of the month, Masud's forces were reported to have outflanked Taliban positions north of the capital and relaunched a massive bombing campaign aimed at the airport northeast of Kabul.

Despite these apparently decisive victories, the position of Dostam remained unclear. A fortnight before his latest victories, on 17 October, he had apparently resumed talks with the Taliban, in his northern stronghold of Mazar-i-Sharif. But had he? The answer was unclear, for other reports at that time claimed that he had pledged support for Rabbani.

14

Collapse of the Soviet Empire (1989-1990)

The Soviet Union's defeat in Afghanistan was not its first but its second strategic reverse in the last decade of its existence. The first, though relatively minor, had occurred on 25 October 1983 when a U.S. force of 1,900 men landed on the small Caribbean island of Grenada. For the first time since the Bolshevik Revolution, a Communist government in a sovereign State had been removed by an outside power's military force.[1]

Until then, other reverses had been temporary. And now, in 1989, Moscow decided to withdraw its forces from Afghanistan, after nine futile years ending in failure; and therefore, politically if not militarily, in defeat. From this major defeat, the Soviet Union never recovered. Four years earlier, Mikhail Gorbachev had taken over from the almost totally inert head of the ruling Politburo, the late Konstantin Chernenko. He soon revealed himself as an outstanding master of international public relations, a useful talent, since his main driving force was to persuade the richer Western powers, especially the United States, to pay for the survival of the dying Soviet Union.[2]

To quote a relevant passage:

The Soviet defeat was remarkable because the patriotic rebels were fighting not only the heavily armed Soviet force, totalling 110,000, but also a Soviet-supplied regular Afghan Army, which was viewed by the guerrillas as puppets. The Soviet-supplied military force was a variable machine, reaching 80,000 at its largest but dwindling to 27,000 through desertions and casualties.[3]

Having decided that there was no point in continuing the War, the Soviet Union completed the withdrawal of its forces before midday on 15 February 1989, as stipulated under the Geneva agreements signed in April 1988 by the Soviet Union, the United States, Afghanistan and Pakistan. The war had lasted nine years, during which the invading forces had suffered 15,000 deaths in action and 37,000 wounded.

According the broad estimates of the U.S. State Department, more than a million Afghans—civilians as well as combatants—had been killed, while around 5 million Afghans had taken refuge in Pakistan and Iran.

Commenting on the end of this ferocious war, shortly after the Soviet withdrawal, I wrote: "In strategic and historical terms, Mikhail Gorbachev's decision to pull the Soviet forces out of their Afghan quagmire constitutes a major reverse for the October Revolution, comparable to the American decision to withdraw from Vietnam in 1973, In each case, the superpower involved tacitly conceded defeat."[4]

On re-reading these words in 2002, I saw no reason to change them, but the passage of 13 years suggests additional comments. America's withdrawal from Vietnam, a major reverse analyzed elsewhere in this book, while it was politically highly damaging to the presidency of Richard Nixon, did not in any sense threaten the survival of the USA; whereas the Soviet retreat from Afghanistan (for this is what it amounted to) can be seen in retrospect as a major cause of the collapse of the Soviet system,

The unstated message reaching the component republics (in effect, the colonies) of the Union of Soviet Socialist Republics was that the Soviet system had ceased to continue its self-imposed task (as Lenin saw it) to spread its system to "all countries of the world, without exception."

When Gorbachev came to power, in 1985, there was no question of his abandoning the stated aims of Lenin, creator of the system in which he had made his career. But he attempted the difficult feat of praising Lenin while not quoting Leninist aphorisms that might inhibit his search for supporters in the non-Leninist countries of the West. That he was conscious of the nationalities issue was clear from the new Program of the CPSU, adopted in March 1986. Indeed, as Robert Bozzo, a distinguished American specialist in international issues, put it: "...the new Program included the proud, albeit specious, declaration that 'the national question, which has remained from the past, has been solved successfully in the Soviet Union."[5]

When Gorbachev replaced the ruling Communist Party of the Kazakhstan Republic's first secretary, Dinmukhamed Kunayev, with a Russian nominee, Gennadi Kolbin, he had broken with convention in a potentially dangerous manner. In Crimea and in the Baltic States (Latvia, Lithuania, and Estonia), in Moldavia and Azerbaijan, in Georgia and elsewhere, nationalist protests broke out.

To be specific, the first real trouble started in February 1988 with street demonstrations in the Nagorno-Karabakh Autonomous Region of the Soviet Republic of Azerbaijan.6 The demonstrators were Armenians, who formed nearly 80 percent of the population: their aim was to be united with the neighboring Republic of Armenia. It was not only an ethnic clash, but a religious one as well, between the Muslims of Azerbaijan and the Christians of Armenia. The clashes went on for three months. Gorbachev restored at least temporary calm by sacking the Communist Party leaders in both the rival republics.

Not long after, a huge demonstration by local Ukrainians took place in Lvov, a town close to the Polish border. They were joined by Armenians and Georgians as well as Lithuanians, Latvians, and Estonians. The activists went still further, by forming a Coordinating Committee of Patriotic Movements of the Peoples of the USSR: a strikingly un-Leninist challenge to the ruling Soviet Party.

Massive demonstrations followed in the three Baltic republics, which of course had been forcibly incorporated into the USSR following Stalin's deal with Hitler. The secret protocols of the Stalin-Hitler Pact emerged into public view when they were, in fact, published in all three of those Republics. Meanwhile, refugees in the tens of thousands were streaming out of both Armenia and Azerbaijan. Long-suffering Armenia in particular was struck by a major earthquake that devastated the Pitak region of the Republic on 8 December 1988: Thousands of lives were lost in this strictly non-political disaster.

The major disturbances chronicled above came three weeks after Gorbachev had faced major ethnic challenges to Moscow's authority. On 13 November thousands had demonstrated in Tbilisi, the Georgian capital, without any attempt at ambiguity: they were protesting against the continuing Russification of their Republic. Three days later, in distant Estonia, the Supreme Soviet, in an emergency session, voted overwhelmingly against Soviet domination. They opted for declaring their Republic sovereign, with the right to veto Soviet laws. Next day, Moscow Radio declared that this decision violated the Constitution of the USSR, and announced that it would be discussed by the Presidium of the USSR Supreme Soviet.

On the 18th, the Lithuanian Supreme Soviet voted to make Lithuanian—instead of Russian—the Republic's official language, and to bring back the pre-war flag and anthem. Encouraging though these measures sounded, they fell short of Estonia's claims, and later that day a crowd of around 100,000 protested at the Lithuanian Republic's failure to vote for sovereignty, in contrast to neighboring Estonia.

On 20 March 1989, the *London Times* reported that two days earlier, Estonian dissidents had pulled down the Soviet flag on their Supreme Soviet building and raised the Estonian national flag. A ban followed on exporting food from Estonia to the rest of the then (still) Soviet Union. In Lithuania, emboldened nationalists called for the withdrawal of the Soviet "occupying forces."

In old times, such multinational protests or acts of defiance would not have happened, or have been brutally suppressed if any such signs of protest had emerged. In these newer times, there were merely a few individual arrests. On 2 March 1989, two Armenians were jailed for eighteen months for taking part in "group activity that violated the social order." A mild central reaction, contrasting with the massacres or mass deportations of the Stalinist and post-Stalinist era.

And then, in Georgia in March and April, the pot boiled over when thousands of citizens of the Abkhaz Autonomous Republic rallied to demand secession from the neighboring Georgian Republic. There was a Georgian counter-demonstration in Tbilisi. Some Georgians went further, staging a hunger strike, laying down tools or stopping trains. The supporting demonstrators numbered around 100,000.

In support of Georgian Interior Ministry troops, Moscow sent in non-Georgian regular soldiers. The demonstrators protested against their arrival with stones and "metal objects." Then a fatal order was given and the regular troops aimed gas canisters at the crowds. Tear gas would have been bad enough, but the gas released was a highly toxic one causing paralysis of the nervous system. The outcome: twenty demonstrators died and 139 were hospitalized.

How would Gorbachev deal with this unprecedented act of rebellion? His well-known foreign minister, Eduard Shevardnadze, was a former first secretary of the Georgian Communist Party. He happened at that time to be on an official visit to Germany. Gorbachev recalled him and sent him to his native country to deal with the crisis in his own way. A subtle man with a capacity for rapid change (and also, possibly, a career Communist who was losing faith in Lenin) Shevardnadze publicly condemned as "intolerable," the deaths of "absolutely innocent people" and put the blame where it belonged: on the local ruling Party.

On 14 April, five days after the massacre, he sacked his local successor, the Georgian party boss, Dzhumbar Patiashvili (who was allowed to "resign"). Some things had changed, but not others, for, in his place, Shevardnadze appointed the local KGB chief, Givi Gumbaridze. Observers of the crisis may have concluded, privately, that the Georgian Party, left to itself, could probably have handled the crisis more skillfully than it did, but were they responsible for the arrival of those non-Georgian regulars armed with fatal gas canisters?

Similar though obviously not identical crises were surfacing in many other areas. For instance, in Tashkent, capital of the Central Asian Republic of Uzbekistan, other demonstrations took place. On 10 and 11 April, following a substantial nationalist rally, troops and tanks restored order on the streets.

In the smaller Moldavian Republic also, thousands were out in protest against Russification: one way of putting the perilous reality of anti-communism. Ethnic protest riots took place in Yakutia, a Turkic-speaking area of northeastern Siberia, as well as in the southern Turkic Republics of Tajikistan and Kazakhstan. Then there were protests in Crimea, whose Tatars had been deported en masse by Stalin (in the same way as he had deported the Chechens) on the ground that they had been collaborating with the invading Germans. Now they were protesting, wanting their former Republic back.

For Gorbachev, the gifted public relations man, these multiple ethnic protests constituted so many challenges in defiance of Leninist doctrine. Not all the problems presented equal perils. Georgia was perhaps easier to deal with than some of the other areas. But unrest in the Baltic Republics constituted a special peril because of their proximity to Poland. As for the Turkic Republics, anti-Russian sentiment was not alone: a complicating factor was the danger of Islamic solidarity beyond the Soviet borders. Moreover, their high birth rates, and therefore their growing percentage within the Soviet armed forces, complicated the problems.

To add to the latter, one need only look beyond the Soviet borders into the new accretions to the wider Communist empire: notably in Poland and Hungary, in East Germany, Czechoslovakia, and Bulgaria. While Albania was still an isolated fortress of Leninist orthodoxy, the same was hardly true of Rumania, where the family tyranny of Nicolae Ceausescu had shown not the slightest interest in such Gorbachevian words as glasnost and perestroika.

To grasp the full range of Gorbachev's nationalist problems after the retreat from Afghanistan, let us look at the problems and challenges of the Soviet Union's newly extended European empire.

At stake was the legitimacy of the Communist Party's monopoly of power; but perhaps even more important were the long dormant but now aroused national aspirations and resentments at what was now openly seen as colonization by Soviet might.

By mid-December 1989, not long after the retreat from Afghanistan, the monopolists were either defeated or actually on the run in the countries listed above: Poland, Hungary, East Germany, Czechoslovakia, and Bulgaria. Even in the Soviet Union, the crucially important Article 6 of the Constitution enshrining the Party's "leading role" was under challenge.

Let us recount earlier signs of trouble.

In Poland, in 1956, the Communist leader, Wladyslaw Gomulka, had defied Khrushchev, despite the presence of Soviet tanks. In Hungary, the Communist leader Imre Nagy had been executed after a mass uprising mercilessly crushed by the Soviet Army. .

In 1981, Poland had come within a hair's breadth of the Hungarian experience. On 13 December 1981, faced with the challenge of the free trade union organization, Solidarity, the authorities had proclaimed martial law. A Military Council for National Salvation was set up, headed by General Wojciech Jaruzelski. At that time, with Brezhnev still in power in Moscow, the Soviet leaders had considered invading Poland, but decided instead to leave Jaruzelski to face the crisis—with much assistance from the KGB and with the Soviet forces ready to intervene under the Warsaw Pact if the situation got out of hand.

On 23 December of that year, President Reagan revealed that the martial law proclamation had been drafted and even printed in Moscow as long ago as the previous September. In that major crisis, Jaruzelski emerged the victor, but not in a long-term sense: Solidarity and its leader Lech Walesa did not consider themselves defeated, in that they looked forward to "fight another day." Indeed, in a round table agreement, significantly on 5 April 1989, at the time of the Soviet retreat from Afghanistan, Wa³esa agreed to an unequal power-sharing deal with Jaruzelski. In other words, the Communists were still in power, but only just. On 4 June that year, a partially free election brought Solidarity an unexpected but sweeping victory: ninety-two of the 100 seats in the newly constituted Senate, and 160 of the 161 seats it had been allowed to contest in the Lower House (the Sjem). A non-Communist coalition took over on the 12th with Solidarity as the senior partner.

In a different way, Hungary went along the same path. On 6 July in that fateful year of the retreat from Afghanistan, Janos Kádar, Khrushchev's choice to restore Moscow's authority in Hungary, died: not before a firing squad but conventionally in hospital. He had, in fact, been ejected in May the previous year as Party boss. His successor. Károly Grosz, was in effect demoted in June 1989 (again, the Afghanistan retreat year) when his formerly all-powerful post as Party boss was devalued. A new general secretary, Rezsoe Nyers, was named as the new party chairman, heading a new four-man presidium.

But the post-Afghanistan revolt went on, as shown on 8 October 1989 when the ruling Hungarian Socialist Workers' Party voted itself out of existence: a party suicide, with denunciations of its Stalinist past.

In East Germany, too, and in Czechoslovakia, the Brezhnev Doctrine was tested to destruction. On 27 October (still in the fateful year) the Warsaw Pact foreign ministers proclaimed the right of each nation freely "to choose the roads of social, political and economic development, with no external interference." As if to proclaim the year of the Afghanistan retreat, the Soviet Union itself officially condemned the Soviet occupation of Czechoslovakia in 1968 as "an interference in the internal affairs of sovereign Czechoslovakia"— in other words, the "raison d'être" of the Doctrine.[7]

Still in 1989, 10 November brought the most dramatic incident in the post-Afghanistan collapse when the Berlin Wall, infamous symbol of Soviet repression, was breached. For weeks, signs of an impending collapse in East Germany had been multiplying. On 7 October, Gorbachev had visited East Germany. Ostensibly, the visit was to mark the fortieth anniversary of East Germany as a Communist state. Apparently oblivious of the possibly negative consequences of his visit, Gorbachev called for changes. In effect, he asked for an East German variant of his own perestroika. How? By ridding the regime of its hard-liners at the top. In effect, his intervention amounted to a briefly deferred kiss of death.

On 18 October, the East German Party boss, Erich Honecker, resigned. Significantly, to appease Gorbachev, he chose as his successor a much younger man, Egon Krenz, not because of his relative youth (he was born in 1937) but because he knew Gorbachev would approve. Not that Krenz had a high reputation. To quote the long-serving head of East German Intelligence, Marcus Wolf: "He was ill equipped for the extreme0ly arduous task he faced."[8]

The Politburo of the ruling Socialist Unity Party fell, and mass resignations from the dreaded security service (the Staatssicherheitdienst, or SSD) followed. The time for mass protests had come. On 4 November, a vast crowd, estimated at around a million, was on the streets of East Berlin. A week later, double that number poured into West Berlin The end had come. The West German chancellor, Helmut Kohl, knew the East German regime was at death's door. From Bonn, he rang the new East German "leader" Egon Krenz who (hardly to Kohl's surprise) did not seem to have realized that "the end was nigh." His words to Kohl were: "Reunification is not on the agenda."

Famous last words. On 6 December, he resigned as Chairman of the Council of State and of the National Defense Council. His short-lived successor Hans Modrow met Kohl in Dresden and signed a joint treaty on "cooperation and good neighborliness." Three further installments brought full and formal reunification to divided Germany. Stage 1 was the signing of a Treaty on 18 May 1990 on "the creation of a monetary, economic and social union." Stage 2 was the signing in East Berlin on 31 August of a Treaty of Unification: a massive document totaling 900 pages. The third date, 3 October 1990 brought the accession of East Germany to the (West) German Federal Republic.

In contrast to months of events in Poland and Hungary, and weeks in East Germany, the collapse of communism took only ten crowded days of demonstrations in Czechoslovakia. Interestingly for the West in particular, the dissident playwright Václav Havel, who had been jailed earlier in the year, played the leading role. In Prague on 10 December, President Husak swore in the first Czechoslovak non-Communist majority government in forty-one years. He then stepped down, having stood for Soviet repression since the near-collapse of 1968.

In Bulgaria, too, events moved fast, even though that country had long been considered virtually an appendage of the USSR. On 10 November, the long-reigning party boss, Todor Zhivkov, was ousted. A month later, his successor, Petar Mladenov, promised free elections and ushered in a new Constitution ruling out the Communist Party 's leading role.

And then, at last, Rumania's delayed turn came. The widely detested and corrupt tyranny of Nicolae Ceausescu was overthrown in a bloody revolt. But the president and his widely despised wife Elena tried but failed to escape. After a secret military trial, both were executed on what, in non-Communist countries, was Christmas Day.

By now, after this extraordinary sequel of events following the Afghanistan fiasco, the Soviet empire had collapsed, leaving only Albania as a self-isolated Stalinist outpost.

Part 4

Three Minimal Wars

15

The Suez Fiasco (1956)

On the international scene, with one obvious exception, no British name was more famous than Anthony Eden. The huge exception was, of course, Winston Churchill. Eden was tall, good-looking in a film stellar way, and his knowledge and experience of foreign affairs were probably unparalleled. During World War I, he had served as an infantry officer. Two of his brothers had been killed; and his own elder son had lost his life in World War II.

His vocation, clearly, was to run international affairs for Britain. Emerging from the First World War, he had studied Oriental languages (Persian and Arabic) at Christ Church, Oxford. As early as 1931, after nearly eight years as a member of Parliament, he was appointed under-secretary of state for foreign affairs. Four years on, in December 1934, he became foreign secretary.

This was a difficult time: Neville Chamberlain had taken over from Stanley Baldwin as prime minister, in May 1937, and lost no time in launching his policy of appeasement. The target of the policy was of course Adolf Hitler. Eden, to his credit, was preaching a policy of collective security and action. The two policies were incompatible, and Eden resigned on 20 February 1938.

Despite his resignation, Chamberlain brought Eden back, as dominions secretary. Eden's career truly blossomed, however, when Churchill took over in May 1940 and named him secretary of state for war. But not for long: he raised Eden to foreign secretary at the end of the year, and Eden served in that post until the end of the war and Churchill's defeat in the general elections that followed.

In more ways than one, the Eden-Churchill partnership was personal as well as political. Churchill had advised King George VI that if he, Churchill, should die, Eden should succeed him. In 1952, Eden would marry a niece of Churchill's, his first marriage having been dissolved. Four years later, he would succeed Churchill as prime minister. And in April 1956, he faced the events that would ruin his career, known as "the Suez crisis."

The man who brought Eden down was the Egyptian dictator, Gamal Abdel Nasser. It would be more accurate, however, to put it differently: Eden brought himself down by challenging Nasser to a military confrontation that he knew not how to win.

Born in 1918, Nasser graduated from the Egyptian Royal Military Academy in 1954. While serving in the Sudan, he formed a conspiratorial alliance with three fellow officers, one of whom, Anwar Sadat, would later succeed him as president. The four young officers set up a secret organization calling itself the Free Officers, whose first aim was to rid Egypt of its lazy, self-indulgent monarch, King Farouk. On 23 July 1952, they achieved their aim, driving the king into exile.

In 1954, Nasser revealed the immensity of his ambitions in a short book entitled *The Philosophy of the Revolution*. In a more idealistic way, it was a kind of *Mein Kampf*. His ambition knew virtually no bounds. Not only would he become the leader of 55 million Arabs, but his future followers would include 224 million Africans and 420 million Muslims.

To Eden who, unlike most Western politicians, could read *The Philosophy of the Revolution* in its original language, its impact was devastating. Here was a dictator whose ambitions, if unchallenged could lead to another world war.

In January 1956, Nasser promulgated a Constitution that turned Egypt into a one-party Islamic welfare state. By then, the Cold War between the Soviet Union and the Western democracies was in full swing, but Nasser profited from both sides by his professed neutrality. He signed a secret contract with Communist Czechoslovakia for war materiel, while persuading Britain and the United States to provide $270 million to fund the first stage of his major project: a proposed High Dam at Aswan.

A major international crisis broke out on 20 July that same year when the American secretary of state, John Foster Dulles, changed his mind and cancelled Washington's offer on the dam. Next day Britain followed the American example. Completely reversing his former pro-Egyptian foreign policy, Dulles charged Nasser of being "nothing but a tin-horn Hitler."*

---

* I interviewed Nasser at his home in Cairo in May 1956, shortly before he nationalized the Suez Canal Company. He denounced Eden as the man who had attempted to get Russia to

On 13 June, the British withdrew their last forces from the Suez Canal Zone, leaving responsibility for defending the Canal against potential aggressors completely in Egyptian hands. The ultimate "no turning back" date was 26 July: the day Nasser nationalized the Suez Canal Company.

Why the British were taken by surprise, indeed shocked, by Nasser's decision, is itself surprising. Nasser's move was surely a natural reaction to Britain's military departure. Eden spelt out his thinking in his memoirs, *Full Circle*: in his eyes, with his memories of appeasement and its consequences, Nasser's seizure of the Company was the equivalent of Hitler's reoccupation of the Rhineland. If Hitler had been stopped then, there would have been no Second World War. Therefore, Nasser must be stopped, if necessary by force.

By this time, Eden wielded the necessary power for a decision of this magnitude, for he was Britain's prime minister. His successor as foreign secretary, Selwyn Lloyd, shared his views on the way to handle Nasser. And encouragingly, despite differing ideologies, those views were also shared by France's Socialist prime minister, Guy Mollet, and his foreign minister, Christian Pineau. They too saw Nasser as a kind of Hitler on the Nile. Moreover, on possibly flimsy evidence, they regarded Nasser as the evil genius behind the Algerian uprising: the major problem facing the French government at that time.[1]

On 27 July—the day after Nasser had seized the Suez Company—Mollet telephoned Eden to suggest that the two powers should collaborate with Israel against Egypt. Unfortunately for the promising Anglo-French alliance, Eden emphatically rejected this suggestion, in fear of hostile repercussions among Britain's friendly Arab governments.

The following day, President Eisenhower of the U.S. sent Robert Murphy as a special envoy to London, where he conferred with Pineau and Lloyd. Later, they were joined by the American secretary of state, John Foster Dulles. The three governments agreed on a harmless move: to call a conference of maritime powers using the Suez Canal.

---

agree to UN manoeuvres to limit arms in the Middle East. He defended his government's decision to recognize Communist China that week because the Chinese Sommunists, not being members of the United Nations, could supply arms to Egypt if Russia stopped sending them. See *BC, the Morning After: A Study of Independence* (Methuen, London, 1963), pp.27-29. Henceforth, *BC, Morning*.

Although agreeing with Dulles on the need for a conference, the British and French wanted action before words. On 5 August, a joint Anglo-French military planning team was set up. Within three days, a plan was ready. The British fleet would depart from Malta on 9 or 10 September and would disembark on Egyptian soil on the 15th or 16th. The plan was given an encouraging name: Operation Musketeer. The joint force would land at Alexandria and advance directly on Cairo.

Despite the lack of British support, the French stuck to their original plan of working with the Israelis. On 1 September, they told Israel about Operation Musketeer. They had pressed on the British a plan for an Israeli ship to be sent through the Suez Canal, as a provocative measure.

Next day brought a discouraging message from America. Eisenhower wrote to Eden, dissociating his administration from any policy of using force. In friendly terms, his letter said: "I am afraid, Anthony, that from his point onward our views on this situation diverge."[2] These were not the words Eden had hoped to read. Privately and publicly committed to his parallel between Hitler and Nasser, Eden spelled the argument out, in some detail, in his reply to Eisenhower. The president came back immediately, arguing, in moderate terms, that Nasser was being built up into a much more important figure than he really was. The best way to deflate Nasser, the president argued, was to take the drama out of the situation. This time, instead of contesting Eisenhower's views, Eden decided simply to keep them to himself.

Clearly, however, Britain's prime minister had been shaken by the president's views and by the secretary of state's unreadiness to use force against Nasser. On the 10th, presumably briefed on Washington's pacific views, Mollet and Pineau flew to London, in the hope of resolving any British doubts on the issue. To their dismay, they found Britain's successor to Churchill shaken by the American attitude.

Although reluctant to drop his plans for military action against Egypt, Eden had given thought to creating an acceptable reason for going ahead. At a conference of maritime powers in London, the idea of launching a Suez Canal Users' Association had been discussed in detail. Eden persuaded the French to join the scheme in the hope that the proposed SCUA would provide Britain and France with an acceptable casus belli. Eden's idea made sense: let SCUA send a ship through the Canal, with orders not to pay dues to Nasser's new Canal Authority, in the hope that Nasser would refuse to allow the ship to continue. To make sure of a casus belli, the ship would be accompanied by allied gunboats so that any Egyptian refusal could be met with force.

Three days later, Dulles called a press conference, to declare that he never envisaged that SCUA would shoot its way through the Canal. In a clear attempt to kill the Anglo-French plan, he went further: "If we are met with force, which we can only overcome by shooting, we don't intend to go into that shooting. Then we intend to send our boats round the Cape."

During this period of argument and cogitation, the French and Israelis met several times at the political and military levels. Until then, Eden had always rejected any French plan for collusion with Israel, but he listened in sympathy when Pineau, on a secret flight to London, spelled out Israel's plans to attack Egypt in the event of coordination with the French and British.

Another week passed. Then on 1 October, Eden cabled Eisenhower, in what was clearly a last attempt to persuade the president to see things his way. In support of his case, he provided a detailed list of Soviet armaments now in Egypt's hands. Yet another week passed, and on the 11th, Eisenhower replied to Eden, agreeing with him that the Soviets were trying to gain a dominant position in the Near East, but stopping short of agreeing with the prime minister on taking action.

While the tension grew over the Suez Canal and Nasser's ambitions, a major crisis was developing in distant Hungary, where a revolt against Soviet-imposed communist rule came to a head in the last ten days of October. From the Anglo-French standpoint, the Hungarian crisis was strategically encouraging, in that it virtually ruled out any Soviet intervention to defend Nasser's Egypt against the Anglo-French threat. A relevant delaying factor was closer at hand: was Israel about to attack Egypt, and if so, when?

On 23 October, Selwyn Lloyd and Patrick Dean, a senior Foreign Office official, flew to Sèvres (on the Seine river) to coordinate plans with the French and Israelis. The British role in this meeting remained secret, but it became clear after the event that Eden should have ordered the British fleet to set off from Malta in anticipation of the Israeli attack, bearing in mind the time needed to reach Egypt by sea, as compared to the immediacy of bombing attacks. It became clear after the event, however, that Eden wanted, whatever the cost, to avoid any appearance of foreknowledge of the Israeli invasion.

The French tried, but failed, to persuade the British to act speedily. On 29 October, the Israeli forces crossed the border into Sinai. Next day, the British and French issued ultimatums to Israel and Egypt to end fighting, pull back from a ten-mile strip along the Canal and allow an Anglo-French occupation of key points.[3] In distant London, the House of Commons heard Eden claim that the aim of the Anglo-French intervention was to separate the combatants. Had this been the real aim, it was clear that it could not be achieved by landing in the Canal Zone, when the combatants were in fact fighting many miles to the east in the Sinai area. Moreover, the Israeli force had defeated the Egyptians a day or so before the British and French forces landed. Eden was thus deprived of a reason, as distinct from a pretext, for a painfully disastrous operation.[4]

During the crisis, the Soviet leader, Bulganin, had threatened massive military intervention on 6 November, to end the Suez fighting, which in fact ended at that time for quite different reasons. Rightly, Eden ignored the Soviet threat, in the light of Soviet involvement in the Hungarian crisis. More relevantly, by then the U.S. was threatening a veto on oil supplies; and Britain, having spent £100 million on its expedition, faced a massive flight from the pound and a withdrawal of capital. Early in November, Britain's chancellor of the exchequer, Harold Macmillan, had asked the International Monetary Fund for a loan of £300 million, which the U.S. declared itself ready to support, but only on condition that Britain should accept an immediate cease-fire.

On 19 November, a midnight communiqué from 10 Downing Street announced that Eden was suffering from "severe overstrain" and would be flying to Jamaica for three weeks' rest On 10 January 1957, he retired and was succeeded by his chancellor of the exchequer, Harold Macmillan.[5]

What, then, were the political consequences of the Suez affair? On the attacking side, there were no victories. Sir Anthony Eden had lost a war he had started and could have won if he had got his timing right. His French ally, Guy Mollet, would have been a co-victor if Eden had left the timing to him. He survived Eden politically, but not by an impressive margin. He resigned on 21 May 1957.

There were only two victors, on opposite sides: One was Israel, which evacuated Egyptian territory on 22 January 1957, except for the Gaza strip and the area of Aqaba. On 1 March, the Israelis announced the withdrawal of its troops from these areas; on the 29th it accepted a UN proposal to erect a mined fence along the Gaza border, on condition that the barrier on the Egyptian side was patrolled by UN troops and not by Egyptians.

Politically, the only true victor was Nasser—the man Eden and Mollet had tried but failed to remove. Relieved of the military threats to his regime, Nasser made a move towards the main target of his *Philosophy of the Revolution* by persuading Syria to join Egypt in a joint state to be known as the "United Arab Republic" (UAR). This first move towards his known target was also his last, for in 1961, Syria withdrew from the UAR. Nasser continued to insist that Egypt should be known as the UAR, but the painful fact, for him, was that this short-lived experiment was the closest he got to the target of his "Philosophy."

16

Before and After the Six-Day War (1967)

The Six-Day War qualifies as a "minimal war" only by its amazingly minimal duration: in other respects—the speed of Israel's offensive and its strategic importance—it qualifies as an extraordinary military and political achievement. It has indeed been described by an objective observer as "one of the most devastating offensives in the history of warfare." Despite its military success, however, it did not bring Israel the respectful peace it had hoped for on the part of the overwhelmingly numerous Arab countries that surrounded it.

Before returning to the military side of this minimum/maximum war, however, let us recount some relevant elements in the lengthy historical background of Palestine. The Holy Land, as it was long known, was holy to conflicting cultures and religions: Judaism, Christianity, and Islam. To the Jews, this Land of Israel (Erez Yisra'el) was the land promised by God to the Jews. To the Christians, it was the scene of Jesus and his Apostles. And to the Muslims, it was a land of places seen as holy associated with the Prophet Muhammad.

Throughout its history of thousands of years, the frontiers of Palestine have fluctuated strikingly. Under kings David and Solomon, the kingdom of Palestine included much of modern Syria and Lebanon. From early prehistoric times, nearly every power in the Middle East has claimed sovereignty over it, including Egypt of the pharaohs, Assyria, Babylonia, and Persia; Alexander the Great's empire, and his many successors: the Ptolemies and Seleucids, the Roman emperors, the Byzantines, the Ummayads, Abbasids, Fatimids; the crusaders; the Ayyubids, Mamluks, and Ottoman Turks. In more ways than one, the First World War brought about decisive changes. In1917 two Zionist leaders in exile, Chaim Weizmann and Nahum Sokolow, visited the then British foreign secretary (from 1916 to 1919) Lord Balfour, and persuaded him to write a letter to Baron Rothschild, head of the famous Jewish banking family. In turn,

Rothschild persuaded him to issue what became known as the Balfour Declaration, which pledged British support for Zionist efforts to turn Palestine into a home for World Jewry. After the war the League of Nations awarded Britain a mandate over the area.

Britain's allies in general approved of this arrangement. Initially, so did Arab elements in Palestine, notably the Amir Feisal, who agreed to the claims of Weizmann, on condition that his own demands elsewhere were met. Unfortunately, they were not, *and any hopes for a peaceful agreement between the Arabs and the Zionists vanished overnight. Years later, Hitler's massacres provoked a mass emigration of German Jews, seeking refuge in Palestine. For a while, the Palestinian Arabs seemed undisturbed, but as the rate of Jewish immigration rose, anti-Jewish feelings mounted and a campaign of terrorism was launched against the British as well as the Jews.

In May 1939, the British responded to Arab terrorism through a White Paper limiting Jewish immigration to an all-time total of a further 75,000, spread over five years. The White Paper also restrained Jewish land purchases. After World War II, the Jewish immigrants turned anti-British. While the Arabs had welcomed the new British restrictions, their earlier indifference towards the Jews turned into massive resentment. Some anti-immigrant ships were sunk, and one was sent back to Germany with the Jewish immigrants on board. Landing permits to others were refused by the British authorities and many escapees were sent to detention camps in Cyprus.

Britain's Labour government, in power after the war, had pledged itself to help the Zionist cause. By then, Jewish terrorist gangs had emerged and gone into action, with considerable success.

On 22 July 1946, Zionist terrorists blew up British headquarters in the King David Hotel in Jerusalem, causing ninety-one deaths. Rival Arab and Zionist conferences later that year resulted in incompatible demands. Meeting in London, the Arab States proposed an Arab-dominated Palestinian State. At Basel, a Zionist Congress, not surprisingly, called for a Jewish state In February 1947 the British government proposed what it considered a reasonable plan of reconciliation: a partition of Palestine between Arab and Jewish zones, under a British trusteeship. Both the Arabs and the Zionists rejected the plan.

By April, the British were losing patience, and referred the Palestine problem to the United Nations. The General Assembly voted in favor of a partition of the country into Jewish and Arab states, with

Jerusalem under a UN trusteeship. This time the Jews accepted the plan, and the Arabs rejected it. On 17 December, the Arab League vowed to use force to stop the proposed division of the Holy Land, and began raids on the Jewish communities. By then, the Jewish terrorist groups were fully organized: especially the Irgun Zvai Leumi, the "Stern Gang" and, on a relatively smaller scale, the Haganah. Thousands of Arabs started to flee the country.

The British decided to withdraw their forces, and on 14 May 1948, their mandate expired. That day, a Jewish provisional government emerged, under David Ben-Gurion, and proclaimed a State of Israel, with Chaim Weizmann as president. Within two days the new state was recognized by both the United States and the Soviet Union.

And so on, year after year, until in June 1967, war broke out between Israel and a coalition of Arab states. The circumstances dictated a speedy war: a protracted one would have faced the Israelis with the long-term problem of its relatively tiny area and small population. Only thirteen miles wide at its narrowest point, Israel's total area was only 14,000 square miles. A protracted war was therefore out of the question. On the military side, it would have enabled the much larger Arab forces to win. With 250,000 men under arms, the Israeli armed forces faced more than twice that number, but only if all the Arab armies were added together, for a total of 525,000. By far, the largest of the Arab armies was Egypt's, with 240,000 men. The total was more than doubled if the lesser Arab armies were counted: 50,000 for Syria and the same for Jordan and Saudi Arabia. Next to Egypt, Algeria had the largest number of men under arms: 60,000; while tiny Kuwait's contribution was only 5,000. With 1,200 tanks, Egypt's force exceeded Israel's 800 by 50 percent; it also boasted of 450 aircraft to Israel's 350.

Numbers alone, however, do not necessarily win a war; moreover, the Arab world was scattered over a vast area, from Egypt in the East to Morocco on the Atlantic. In a protracted war, the Algerian contingents might have proved useful: another reason for Israel's plan for a speedy strike and early victory. Another factor in Israel's decision to strike first was the awareness that a substantial portion of Egypt's military machine was engaged in Yemen.

The backbone of the air force came from France; impartially, Britain supplied tanks to Jordan as well as to Israel On 30 May 1967, President Nasser of Egypt and King Hussein of Jordan signed a defense treaty. A fortnight earlier, on 16 May, Egypt had demanded the withdrawal of the UN Emergency Force from Sinai—he contested territory separating Egypt from Israel. Over the next few days, five Egyptian infantry divisions and two armored formations entered Sinai; and Egypt had closed the Strait of Tiran to Israeli shipping.

Meanwhile, Israel had been mobilizing its forces, most of whom were highly trained civilians. It has to be understood that, by the nature of its situation, not least its relatively small armed forces, its limited manpower and financial resources, it had no choice other than a short but decisive war. Any conflict longer than two weeks would inflict massive damage on its flourishing economy. A lightning war was therefore the only valid choice.[2]

As Willmott put it, although the war that followed became known as the Six-Day War, in fact victory was achieved in about three hours, in which time, "the Israelis inflicted a colossal defeat on the Arabs from which there could be no recovery." As in previous and later conflicts, the real winner of the war was General Moshe Dayan, who had lost an eye in a Palestinian-Jewish clash with the Vichy French in Syria in World War II. For the rest of his life, his black eye patch became a symbol of military leadership. Although the military plan had been drawn up by the Israeli Chief of Staff, General Yitzhak Rabin, Dayan was regarded by his compatriots as the real architect of victory.

Egypt had lost about 800 tanks, 300 of which had been taken intact. The Israelis had also seized 450 pieces of artillery and 10,000 vehicles. In Palestine itself, they had taken Old Jerusalem, long claimed by the Arabs as a holy Islamic place. From the Jordanians, the Israelis had taken most of their 200 tanks and more than half of their personnel carriers.

The speed of the Israelis' victory had led them to assume that henceforth their domination would be recognized by the Arabs, but this was a serious misreading of the bitterness the Arab world would conserve, along with a thirst for revenge. It was also an exception to the long-standing rule that the victors in a major war can expect, and normally receive, bitter but lasting respect from the vanquished. It soon became clear, indeed, that the reason for this disappointing sequel to a stunning victory simply reflected the fact that—unlike the crushing victories of major powers in coalition against other major powers—Israel's was the widely unexpected triumph of a tiny power crushingly outnumbered by its enemies.

On 4 June, the day before the war broke out, President Aref of Iraq had opened a conference of the Arab oil-producing States in Baghdad. Apart from the Iraqi hosts, the participants were Saudi Arabia, Libya, Kuwait, Algeria, the United Arab Republic, Bahrain, Qatar and Abu Dhabi. On 5 June—the day of the Israeli attack, the conference decided to cut off oil supplies to any state giving aid to Israel. That day also brought anti-U.S. and anti-U.K. riots in Algeria, where thousands of mainly teenage demonstrators sacked the U.S. cultural centre and the British Council Library in Algiers. Anti-British and anti-American rioting also broke out in all other oil-producing countries in the Islamic States of the near East, North, and West Africa, along with a breaking of diplomatic relations.

In the face of these hostile outbursts, the U.S. declared its neutrality on 5 June, the day the Israeli assault had begun. As the State Department put it, the United States was "neutral in thought, word and deed." A White House statement introduced an emotional note, declaring that the U.S. was "deeply distressed" at the outbreak of hostilities, and added that "tragic consequences will flow from this needless and destructive struggle if the fighting does not cease immediately." Indeed, the statement went on, the United States would devote all its energies "to bring about an end to the fighting and a new beginning of programs to assure the peace and development of the entire area."

Next day, the secretary of state, Dean Rusk, described as "utterly and wholly false" an Egyptian allegation that American carrier-based planes were helping Israel. He went on to describe the allegation as "a malicious charge known to be false and therefore obviously invented for some purpose not fully disclosed."

The following day, 8 June, the White House disclosed that a "hot line" between Washington and Moscow had been used for exchanges between President Lyndon Johnson and the Soviet leader, Alexei Kosygin.

An Arab leader who had found it particularly difficult to survive the humiliating political defeat was President Nasser of Egypt. On the 9th, he announced that he had decided to give up all official and private functions and become a private citizen. He had asked a former prime minister, Zakaria Mohieddin, to take over the Presidency of the Republic. He went on to claim that U.S. aircraft had "reconnoiterd our positions," while British aircraft had raided positions on the Syrian and Egyptian fronts. He went on to give a detailed outline of the events that had led up to the war and to recall the acts that had led to the recovery of the Suez Canal and the building of the famous High Dam "to turn the arid desert green."

His decision to retire as president could hardly have been more short-lived, however. Evidences of his personal popularity were pouring in. Hundreds of thousands of people, many of them in tears, crowded the streets of Cairo and Alexandria to plead with Nasser not to quit. Later that day, Nasser made another statement, broadcast by Cairo radio, saying he had been "profoundly touched" by the popular demonstrations in his favor. Next day, sixteen hours after his initial broadcast, he announced that he had decided to remain in office "in view of the people's determination to refuse my resignation." He went on: "Therefore I have decided to stay where the people want me to stay, until all traces of aggression are erased."

On 5 June, the day the war broke out, the Soviet government had issued a statement condemning what it described as Israel's "aggression." The statement included a threat, which the speed of Israel's victory rendered unrealistic, by reserving the right "to take all steps that may be necessitated by the situation." Having called on Israel to cease "immediately and unconditionally its military activities against the UAR, Syria, Jordan and other Arab countries," and to withdraw immediately, Moscow's statement accused Israel of pursuing a course of "recklessness and adventurism." The statement also alleged that the Israelis had started the attack with "encouragement by covert and overt actions by certain imperialist circles." This was classical "Soviet-speak," without any factual backing.

Unusually, the Soviet attack was supported by all the East European "countries, with one exception: Rumania, whose media reported the crisis factually and objectively. The other Communist countries —the Soviet Union, Bulgaria, Czechoslovakia, East Germany, Hungary, Poland, Yugoslavia—ended a meeting in Moscow with a statement denouncing Israel's "aggression and expressing support for the Arab countries. Rumania did take part in the meeting, but declined to sign the statement.

On 10 June, the Soviet Union broke off diplomatic relations with Israel; and the other Communist countries followed suits—again, with the sole exception of Rumania. Simultaneously, the UN Security Council met almost continuously to discuss the crisis, and specifically a Soviet-drafted resolution denouncing Israel's "aggression" and calling on Israel to withdraw its troops "immediately and unconditionally." Interestingly, the Soviet resolution received only three supporting votes (from Bulgaria, India, and Mali).

Two days later (12 June), the prime minister of Israel, Levi Eshkol, addressed the Knesset (Parliament) in firm though emotive terms. He made it clear that he was addressing, not only the Knesset, but "all the nations of the world." There should be "no illusion," he declared, "that the State of Israel is prepared to return to the situation that reigned up to a week ago.... Alone we fought for our existence and our security. We are entitled to determine what are the true and vital interests of our country and how they shall be secured. The position that existed up till now will never again return. The Land of Israel will no longer be a No Man's Land, wide open to acts of sabotage and murder..."

He went on to accuse the United Nations of having failed to condemn nineteen years of Arab hostility towards Israel—a situation he described as unique and unparalleled in international relations.

On 17 June, an emergency session of the UN General Assembly voted a resolution rejecting the unification of Arab Jerusalem. Thirteen days later, the Israeli government proclaimed the union of both parts of Jerusalem, in defiance of the UN resolution. At the end of August, Arab leaders (not including Syria, Algeria, and Tunisia) met in Khartoum (Sudan) and debated how to proceed with a draft common position: "No peace with Israel, no recognition of Israel, no negotiation with it and insistence on the rights of the Palestinian people in their own country." Thus was proclaimed a war without end.

On 4 February 1969, Palestinian guerrilla organizations met in Cairo and chose Yasser Arafat as chairman of the Palestine Liberation Organization. Born in Jerusalem in 1929, Arafat was educated in Cairo, where he took a degree in civil engineering and became the leader of the Palestinian students at Cairo University. At twenty, he was running guns for the Arab side in the clashes of 1948, which led to the creation of Israel and the exodus of Palestinians. During the prolonged troubles of 1951-52, which drove the British out of the Suez Canal Zone, he was both training and leading the Palestinian and Egyptian commandos who harassed the British. He gained further battle experience as a commissioned lieutenant in Nasser's Army during the Suez crisis of 1956.

Two years later, he moved to Kuwait, where he edited a nationalist magazine, Our Palestine. Under the pseudonym of Abu Amar, he joined the organization of which he became the leader, Al Fatah, the name of which deserves dissection. It is a kind of acronym for its full title, Harakat al-Tahrir el-Watani al-Falastini (Movement for the Liberation of Palestine). In Arabic, the initials (HTF) mean death. Reversed, (into"fath"), they spell conquest.[3]

When these lines were written, twenty-three years later, Arafat was under constant siege and permanent threat of death, from the Israeli forces under the political control of Major General Ariel Sharon, the bellicose prime minister of Israel. Of minor interest, perhaps is the fact that Sharon and Arafat were almost exact contemporaries, born in 1928 (although it is sometime claimed that Arafat's real year of birth was 1929.

When these lines were written (in November 2002) there was no credible sign of a settlement—short or long term—of the Israel-Palestine problem. A new *intifada* (uprising) had been launched in October 2000. In December of that year, the then U.S. president Clinton had submitted a draft peace accord to Arafat and the Israeli prime minister, Ehud Barak had accepted the plan, not of course as drafted, but as a basis for further negotiations between the two sides. As for Arafat, he had raised a list of twenty-five separate questions for discussion.

Although Israeli and Palestinian negotiators met in Washington on 4 January 2001, they did discuss possible ways of reducing the current violence, but avoided any substantive discussion of the U.S. proposals.

In a special premiership election held in Israel on 6 February 2001, Ehud Barak (who had called it) was swept out by Ariel Sharon, by now rightly regarded in Israel and abroad as a leading right-wing hawk. Any random search during the forthcoming months merely confirmed the continuing trend: any apparent agreement for a truce between the warring sides would be short-lived. For example, a Palestinian-Israeli cease-fire was agreed on 13 June 2001, but began to collapse in the first week of July. Thereafter, scarcely a week passed without acts of violence by each side against the other. The devastating suicide attack on the New York World Trade Center on 11 September that year brought televised scenes of anti-U.S. jubilation in the Palestinian self-rule areas. To be fair, Arafat and other Palestinian officials did condemn the suicide bombings. I know of no evidence that they were in any way involved with the extremist Al-Qaeda international terrorist network.

Arafat and Israel's (relatively moderate) deputy prime minister. Shimon Peres met for talks in the Gaza Strip on 26 September, and agreed to reactivate the earlier cease-fire agreed in June. Almost predictably, new clashes erupted on the Gaza Strip the following day. In the face of these repetitive clashes. President George W. Bush was moving away from what the Palestinians regarded as a pro-Israeli attitude. Importantly, on 2 October, Bush publicly (at a press conference) affirmed U.S. support for an independent Palestinian state next to Israel. In an address to the UN General Assembly in New York on 10 November, the president declared that the USA was "working toward the day when two States—Israel and Palestine— live together within secure and recognized borders as called for by the Security Council resolutions."

Personally, as the author of this book, I have long held exactly this view. Indeed, it would make sense if this sensible and logical solution could have been achieved before the threatened U.S. attack on the Saddam Hussein regime in Iraq is launched. When these lines were written (in late November 2002), there was still no visible sign that an independent Palestine State would be in existence before a U.S.-British assault on Iraq was launched. Without an independent Palestine, a Western assault on Iraq would cause a mounting anti-Western virulence to burst out, even from Islamic States with no particular liking for the Iraqi dictator.

17

The Falklands Case (1982)

The best-known president of Argentina, internationally, was undoubtedly Juán Perón. Perhaps the least-known was Dr. Arturo Illia, whose Popular Radical Party swept the Peronists out of his way in the elections of May 1963. Two months later, he was elected president.

As it happens, President Illia received me in Buenos Aires, in August 1965, in the huge presidential office of the Palacio Rosado. A friendly, informal man, he answered my questions, then told me he would like to brief me on the Islas Malvinas, better known to British visitors as the Falkland Islands Dependencies. In retrospect, his short lecture can be seen as a friendly, informal warning that Argentina was serious in its claims to sovereignty over the disputed islands.

Pointing to a wall map, he drew my attention to the area known as the Argentine Antarctic, which, he said, was a natural extension of the continental shelf, of which the Malvinas were a part.

Although I had listened with interest, I was not particularly impressed by the new president's words. An unusual peculiarity of the islands was that their inhabitants numbered only about 2,000 (with a more imposing animal number of nearly 600,000 sheep). Most of the inhabitants were of British origin, but they included a considerable proportion of Scandinavian ancestry. The islands had been under successive occupations: the Spaniards, with support from Buenos Aires, had bought out the French and expelled the British. In 1833, however, the British expelled the Spanish speakers, including Argentine settlers. Ever since, the Argentinos had nursed their grievance.

On my return to Britain in late 1964, I wrote an account of my two-months tour, entitled "Latin American Journey," for the magazine *Encounter*, which included the following passage:

The humans are not, as the Argentinos say, a transient population of temporary migrants: 80 percent were born there and have no wish to come under Argentine rule. Nor are they an oppressed colony: the "colonialism" issue is absurd; there is no national liberation movement in the Falklands. On the other had, do we really still want the Falklands, Gibraltar and other scattered bits of real estate that once seemed so essential? Is there not a basis here for long-term bargaining: a delayed transfer of sovereignty in return for solid economic concessions, for instance special rights for British fishing?

The Argentine president may have hoped that I would report our conversation to the Foreign Office (or to MI-6), but I did not, partly because I believed there were far more important matters to report, and partly because at that time Argentina did not appear, in any sense, to represent a threat to British interests.

Eighteen years later, in 1982, a new sense of urgency reigned, on the Argentine side.[1] The Falklands question dominated diplomatic talks in New York on 26 and 27 February between the UN representatives of Britain, and Argentina. Britain's team was headed by Richard Luce, minister of state and the Argentinian by Enrique Ros, deputy foreign minister. Also present were two Falklands representatives: John Cheek of the islands Legislative Council, and Tony Blake, a farmer.

The two-days' talks illustrated a situation more visible to outsiders than to participants reluctant to recognize realities. A joint communiqué at the end of the talks stated: "The meeting took place in a cordial and positive spirit. The two sides reaffirmed their resolve to find a solution to the sovereignty dispute and considered in detail an Argentine proposal for procedures to make better progress in this sense."

The "joint" statement was clearly better designed to keep the British happy than to meet Argentinian ambitions. A mere two days after the talks ended, the Argentine Foreign Ministry issued a statement in Buenos Aires declaring that Argentina had negotiated with Britain for more than fifteen years "with patience and good faith" under the terms of UN resolutions. But unless a speedy negotiated settlement was reached, Argentina would "put an end" to negotiations and "seek other means" to resolve the dispute.

During question time in Britain's House of Commons or 3 March (two days after the threatening Argentine statement), Richard Luce reaffirmed that Britain had no doubt about its sovereignty over the Falkland Islands or about its duty to the Falklanders, The Argentine statement had caused "deep anxiety" and "there can be no contemplation of any transfer of sovereignty without consulting the wishes of the islanders, nor without the consent of this House."

Sovereignty was of course the key issue. The previous year, the Argentine government had issued a six-point declaration stating that any realistic negotiation on the Falklands "must presuppose Argentinian sovereignty over the islands, as an essential point for reaching a solution." Not surprisingly, reports of the declaration appeared in the Argentine press on 27 July 1981.

An important role in the developing crisis was played by an American high official, Thomas Enders, assistant secretary of state for Latin American affairs. Despite this title, which had been conferred on him lately,

Enders made no claim to expertise on Latin America. His recent assignments had taken him to Yugoslavia and Cambodia. His imposing height (six feet eight inches) and his natural "can do" approach to difficult problems brought him words of praise from his boss, Secretary of State Alexander Haig, but made enemies and aroused irritation among America's British allies.[2]

In retrospect, it is fair to say that Luce and Enders played the diplomatic game by the accepted rules: that is, each in support of his country's interests. As it happened, America's interests, at that time, favored Buenos Aires, in that the U.S. aimed at encouraging Argentina to become a politically stable ally, with a sound economic base. In comparison, Britain's sovereignty over the Falkland Islands was of minor importance. According to Britain's *Sunday Times* book on the Falklands War, already quoted, Enders at that time "scarcely knew where they [the Islands] were."[3]

Understandably, when Luce called on Enders at the State Department and urged him to encourage the Argentinians to keep on talking, Enders responded by asking his visitor whether the British were interested in a negotiated solution. Luce said they were: he could hardly have said anything else. Smiling, Enders said he would pass the message on to Buenos Aires. Coming from the mighty United States, the message could only encourage the Argentinos to believe that, in due course, they would persuade the British to transfer their sovereignty over the Islands.

Certainly, as Enders had presumed, this was how the Argentinos interpreted his message. And when, two days later, he received a brief telegram from Britain's foreign secretary, Lord Carrington, confirming his approval, Enders felt free to present his interpretation of his talks with Richard Luce to the Argentinian foreign minister, Costa Mendes. This he did directly, when the two met *à deux* in Buenos Aires.

What neither the British nor the Americans apparently knew, at that time, was that it was already too late for diplomatic manoeuvres. In early December, Argentina had already decided to prepare to invade the Malvinas, if Britain continued to claim British sovereignty over them. This bold decision had been taken in early December, in secret, by President Leopoldo Galtieri and Admiral Jorge Anaya, commander of the Argentine navy.

By late March, it was clear that Argentina was about to attempt naval and military action in the Falklands, although the degree of the force contemplated was not clear. There were unofficial reports that an Argentinian aircraft carrier and two destroyers were on their way south. Meanwhile, the British nuclear-powered submarine *Superb* had left Gibraltar on an unstated mission. On 30 March, Lord Carrington, back in London from a European Council meeting in Brussels, made a fresh call for a diplomatic solution but avoided spelling out Britain's defensive measures. In a House of Commons debate that evening, the opposition Labour Party's spokesman on foreign affairs, Denis Healey, called the government's handling of the dispute "foolish and spineless," "Recent events," he thundered, "have found the government with their trousers down in the South Atlantic."

No form of words could have been better chosen to encourage the Argentinians to go ahead. They did: in the early hours of 2 April, when a contingent of their troops overwhelmed the seventy Royal Marines stationed on the Islands in 3 1/2 hours. The British Governor, Rex Hunt, was deported to Uruguay. Back in Britain on the 5th, he confirmed that there had been no British casualties, while five Argentinians had been killed and seventeen injured. Argentina's governing military junta announced that "the armed forces today recovered the Malvinas, the Georgias and the South Sandwich Islands for the nation." The statement went on to call for a collective effort to "convert into reality the legitimate rights of the Argentinian people which had been patiently and prudently deferred for almost one hundred and fifty years."

Next day—2 April—President Galtieri addressed the nation, on a triumphal note. The decision to recover the islands, he said, "was prompted by the need to put an end to the interminable succession of evasive and dilatory tactics used by Great Britain to perpetuate its dominion over the islands."

For Britain's first lady prime minister, Margaret Thatcher, this was an opportunity not to be missed. Internationally, as well as nationally, the response to the Argentinian invasion was overwhelmingly critical. On 3 April, the UN Security Council had demanded an immediate withdrawal of all Argentine forces. The majority in favor had been ten to one, with Panama the only country voting against. The majority included the United States and France. The abstentions, however, had included Spain (for reasons reflecting its long history of involvement in Latin America) as well as the three Communist members: the Soviet Union, China, and Poland.

The Council, however, had called on Britain and Argentina to seek a diplomatic solution to their difficulties.

Despite the Security Council voting, Galtieri on 4 April had defended its decision to act by denouncing Britain for illegally occupying the Malvinas since 1833. He went on: "If the Argentinian people are attacked by air, sea or land, the nation in arms will go to battle with all means at its disposal. We will not withdraw from Argentinian territory. Argentina will maintain its freedom, to protect the nation's interest and honour. It will not be negotiated."

By then, within hours of the invasion, the U.S. State Department had issued an announcement deploring the use of force and calling on Argentina to withdraw its troops.

On the 3rd, the foreign ministers of the European Community had jointly condemned the invasion and called for the immediate withdrawal of the Argentinian troops and compliance with the Security Council's call to seek a diplomatic solution.

That same day, Britain's House of Commons held a three-hour emergency debate on the crisis. In her opening speech, Mrs. Thatcher noted that for the first time in many years, "British territory has been invaded by a foreign power." The lawful British government of the islands had been usurped. She called on the House to join her in "condemning totally this unprovoked aggression by the government of Argentina against British territory: It has not a shred of justification and not a scrap of legality."

In the debate that followed, however, the government found itself at the receiving end of a considerable wave of criticism, which came not only from the Opposition Labour and Social Democratic parties, but even from individuals within the ruling Conservative Party. These included Edward du Cann, chairman of the backbenchers' "1922 Committee" and Enoch Powell, the maverick MP renowned for his erudition and outspoken views. Thus Du Cann: It was astounding that Britain appeared to have been so "woefully ill-prepared," despite its "huge" defense expenditure. The assumption that the problem could be solved by diplomatic means alone was "surely fatuous."

Powell: Mrs. Thatcher had "received a soubriquet as the Iron Lady" (in a speech in 1974 about defense against the Soviet Union and its allies), and in the next week or two, the House, the nation and the prime minister herself would "learn of what metal she is made."

Michael Foot, leader of the main Opposition Labour Party, declared that the people of the Falklands had "the absolute right to look to Britain at this moment of their desperate plight." The Falklanders had been betrayed and the government must now prove by deeds that they were not responsible for the betrayal.

David Owen questioned why no action had been taken amid the ample warnings that the position was deteriorating.

In contrast to his Parliamentary friends and enemies, the defence secretary, John Nott, played a commendably firm role, making sure that the armed forces, especially the Royal Navy, would be well armed and prepared for the distant clash of arms that awaited them. For his positive attitude, he was praised by the chief of the defence staff, General (later Field-Marshal Lord) Bramall.[4]

Considerably less determined in the face of criticism was the foreign secretary, Lord Carrington, who resigned the following day (5 April), accepting responsibility for the conduct of government policy on the Falkland Islands and acknowledging that the Argentinian invasion had been a "humiliating affront to this country." He was replaced by Francis Pym, the leader of the House of Commons.

On 9 April, the Defence Ministry issued a list of the huge British naval task force—nearly seventy ships—on its way to the Falklands. It included the carriers *Hermes* and *Invincible*; the destroyers *Antrim* and *Glamorgan*, various frigates and the assault ships *Fearless* and *Intrepid* and nuclear-powered hunter-killer submarines. The task force also included between 5,000 and 6,000 troops and between 2,000 and 3,000 Royal Marines.

The Argentinian forces too were by no means inconsiderable. They included the aircraft carrier *25 de Mayo*; the cruiser *General Belgrano* (later sunk); various frigates, corvettes and submarines; and sixty-eight Skyhawk fighter-bombers.

After considerable fighting in and around the Islands in May and early June, the Argentinian forces surrendered to the British task force on 14 June. It had been a costly war, in money rather than lives. The financial cost to Britain was estimated at around £700 million; in British lives, 254, against 750 Argentine deaths. On 17 June, three days after the Argentine surrender, General Leopoldo Galtieri resigned as president of Argentina and as commander-in-chief of its armed forces. He complained that the Army had failed to provide him with the political support to continue in office. Two days earlier mass demonstrations outside the presidential palace had blamed him for "selling out" his country. "No surrender!" they shouted at the empty balcony of the Casa Rosada. There were no words of sympathy, only insults, such as *"Galtieri, hijo de puta!"* ("son of a whore").[5]

In contrast, and not surprisingly, Margaret Thatcher's popularity had soared, reaching an unprecedented 80 percent in the public opinion polls. Once more, the popular label of the "Iron Lady" reigned in people's minds. It was not, of course, a permanent label for the great majority of the electorate, but in some respects, it was deserved. The Iron Lady had stood up against an invasion ordered by a military dictator many miles away. The dictatorship had lost; democracy had won.

Part 5

Tyrant Wars

18

Saddam Hussein's First War (1980)

The Arab world is not noted for its attachment to democracy, which is virtually non-existent in all the Arab countries, from Morocco in the Atlantic West to Saudi Arabia and the Sultanates in the Middle East. Not all the Arab dictators are evil tyrants, however: a worthy exception was the late King Hussein of Jordan, who held absolute power but did not abuse it. In a roll call of Arab tyrants, Saddam Hussein of Iraq would earn the highest (or lowest) rank. During his turbulent reign, he launched two "Gulf Wars": in 1980 and 1990. Both ended in disaster for him; yet he remained in power, by virtue of his ruthless tyranny.

Turbulence had long reigned in oil-rich Iraq. The young King Faisal II had acceded to the throne on his eighteenth birthday on 2 May 1953. Five years later, on 14 July 1958, he was assassinated in an Army coup led by General Abdul Karim al-Kassem, who proclaimed a Republic. He claimed Kuwait (which had become an oil-rich principality) as an "integral part" of Iraq. His tenure of power ended on 8 February 1963; and his life ended that same day. Turbulence continued under the new regime, initially controlled by Nasserites, as supporters of Egypt's president Gamal Abdel Nasser were known.

In the shadows, however, the authoritarian Arab Socialist Ba'ath Party was biding its time. The key word was "Ba'ath," meaning "Renaissance." Its founders were two Syrians: Michel Aflaq and Salah al-Bitar. Their ideology was relatively secular: a combination of socialism with Arab unity. Its Iraqi leaders included the future dictator Saddam Hussein.

The Ba'athists played a minor role in the overthrow of the military dictator Kassem. In the early 1970s, Saddam and his senior colleague Ahmad Hassan al-Bakr dominated the Ba'ath party. On 16 July 1979, however, Bakr resigned for health reasons, and Saddam Hussein became the sole leader of the party. At any rate, that was the visible state of affairs. But only two weeks later, the Ba'ath's military wing, taking advantage of Bakr's departure, tried to seize power. This was Saddam's opportunity. He simply took over. A fortnight later, an alleged rival plot to seize power came to his attention. On 8 August 21, so-called "plotters" were executed on his orders. Among them were five members of the Revolutionary Command Council and two of his ministers.[1]

Over the next few weeks, Saddam Hussein appointed relatives and friends of his in his native town, Tikrit, to all key posts. His one problem in terms of power was the fact that he had never served in the Army. He overcame this potential handicap by raising the salaries of service officers and reserving special benefits for lower ranks. His regime of terror became absolute, as he would demonstrate on occasion by personally shooting potential rivals in his own presidential office.

A minor but not irrelevant incident is worth recalling. Saddam's elder son Udai had killed his resident kitchen "taster" in a fit of rage. Saddam allowed him to be jailed, but only briefly. After all, like father like son....

Saddam had achieved absolute power, but he craved international prestige, and in September 1980 he took a disastrous decision: to invade neighboring Iran. He had reckoned on an easy victory. Some months the Iranian Islamic leader, Ayatollah Ruhollah Khomeini, had driven the ruling Shah, Reza Pahlavi, into exile. The Imperial Army appeared to be disorganized and weakened by a sweeping purge, and Saddam envisaged a quick victory.

Both countries had inherited an imperial past. Iraq had been part of the former Turkish Empire; Iran had in effect succeeded the Persian Empire. There was a long-standing border dispute between the two countries centered on the area now known as Khuzestan. This contested area contained most of Iran's oil reserves, although its original population was overwhelmingly Arab.

In 1975, the then shah of Iran (who was expelled four years later) obtained border concessions from Iraq under the Treaty of Algiers. In return, the shah pledged to end his country's backing for the Kurdish guerrillas in Iraq, who had been fighting for independence in the northeast for several years. The "peace" between the two countries might have been long-lasting, but for a major event in Iran in 1979, when the Shah was expelled and a fundamentalist Islamic regime took over.[1]

There was an aggravating detail: the new regime was controlled by the long-exiled religious leader, Ayatollah Khomeini, a fanatical member of the Shia version of Islam. Although the Shiites formed a majority in Iraq, Saddam Hussein's ruling clique belonged to the rival Sunni sect. Initially, the post-shah situation in Iran was somewhat chaotic.

Saddam Hussein weighed up the situation and concluded that this was a good time to launch a war against Iran. Although he had negotiated, and signed, the Algiers treaty, its terms rankled. He had several reasons (or pretexts) to attack Iraq's non-Arab neighbor. One was the Shah's seizure of three small islands in the mouth of the Persian Gulf: Abu Musa and the Greater and Lesser Tumba. More important, no doubt, from Saddam Hussein's viewpoint, was that victory in a war with Iran would probably enable him to "displace" Anwar Sadat of Egypt as the accepted leader of the Arab world.[2]

On paper, the relative balance of power between Iraq and Iran appeared to be more or less even. The Ayatollah's regime had inherited impressive forces from the ousted shah, including 875 Chieftain tanks, more than 800 U.S. M48 and M60 tanks and more than 600 helicopters. The air force disposed of 188 F-4 Phantoms and 166 F-14 Tomcats. The navy was the largest and most modern in the Gulf. The statistical numbers were deceptive, however, in that the seizure of the American embassy in Tehran in November 1979 had in effect demobilized two-thirds of the Army and Air Force reliant on U.S. maintenance of their weapons.

Until that incident, Iraq and Iran in effect mirrored the Cold War, for while the Iranian equipment had come from America, the Iraqi Army was heavily dependent on Soviet deliveries. The Iraqi army had 2,500 Y-54 and T-62 tanks and around 1,000 field guns. The air force consisted mainly of Soviet-provided MiG-21s and MiG-23s. Border skirmishes grew steadily worse in 1980, culminating on 4 September when the Iranians attacked two border villages on the Baghdad road.

For Saddam, this was the turning point, and he ordered his planned offensive against Iran on 12 September. The Iraqi ground forces made rapid gains in the south, while the air force bombed Tehran airport and other targets. By all accounts, Saddam Hussein, basking in the success of these initial attacks, allowed himself to suppose that Iran's apparently unstable new regime would aim at a peace settlement. He was soon disabused. The Iraqi attack had, at least temporarily, united the Iranians. Even in the contested territory of Khuzestan, they rallied against the Iraqi invaders.

There is no need, in the context of this work, to recount the war in all its details. Saddam Hussein and his generals had undoubtedly under-estimated the Iranian armed forces and the determination of the people under their new regime. Far from being weakened by the invasion, the new Iranian government was strengthened. As Michael Orr—a recognized expert—puts it, "...the war quickly took on a peculiar character, with neither side making an all-out effort. The Iraqis were content to bombard the cities from a distance and seemed in no hurry to begin fighting within the cities or even to isolate them completely....The Iranians did not try to launch major counter-attacks...It is possible that the Gulf War was the first of a new form of war in the Third World, where countries have been supplied with sophisticated equipment which it is beyond their skill to operate."[3]

The war dragged on for years, without any sign of serious strategic planning on either side. During the first few months of 1987, Iranian spokesmen called repeatedly for the overthrow of Saddam Hussein, the replacement of Iraq's Ba'athist government and the punishment of Iraq as the "aggressor." By that time, Saddam had evidently shelved his earlier optimism for ousting Iran's government. His spokesmen were declaring repeatedly that they were willing to pull out all their forces to the international frontier and would agree to a cease-fire. By then, Iraqi forces had ceased to make further advances and indeed were mainly defending their country's territory against repeated Iranian assaults.

The U.S. permanent representative at the United Nations, General Vernon Walters, visited Moscow and Beijing between 30 June and 4 July to secure an agreement on ending the war. After considerable resistance on Moscow's side, a Soviet spokesman declared that the Soviet Union would "undoubtedly" withdraw its warships from the Gulf if the U.S. also agreed to pull out. On 20 July, the UN Security Council adopted Resolution 598 demanding an immediate cease-fire and the withdrawal of all forces to international boundaries.

The Iraqi government welcomed the resolution and declared its willingness to comply with its terms but only on condition that Iran did likewise. Having, in effect, almost defeated the Iraqi invaders, the Iranian representatives offered themselves the luxury of sardonically critical anti-American statements. In a speech to the UN General Assembly on 22 September, President Khamenei (not to be confused with Ayatollah Khomein) described Resolution 598 as "an indecent, condemnable position" forced on the Security Council "by the will of the big powers, particularly the United States." He described the Security Council as "a paper factory for issuing worthless and ineffective orders." Nevertheless, he stopped short of rejecting the resolution.

On 7 October, Iran formally broke off diplomatic relations with Iraq; somewhat late in the day, seven years after the outbreak of fighting. The war dragged on, in fact, until 18 July 1988 when Iran announced its acceptance of Resolution 598 "in the interests of security and on the basis of justice."

In part, no doubt, because of this choice of words, Iraq shortly afterwards carried out a series of major air attacks on Iranian targets. These culminated on 1 August 1988, when Iraqi planes dropped chemical bombs on economic targets in Western Azerbaijan, wounding more than 2,400 people. (Already, on 16 March of that year, 1988, Saddam had tried out his chemical weapons to massacre thousands of Kurds who, in Iraq as well as in Turkey, formed a defenseless but unaccepted minority.) Despite repeated warnings by Iraq that it would not accept an "imposed" cease-fire date, the UN secretary-general, Javier Pérez de Cuella, announced on 7 August that a cease-fire would take effect on 20 August.

Both sides announced their acceptance of that date. Iraq staged national celebrations to mark its "victory" in the war, despite the inescapable evidence that it had resulted in a stalemate—that is, a victory for Iran.

Despite this encouraging announcement, the "peace" negotiations dragged on for months. Face-to-face talks had opened on 25 August 1988. Each negotiating team was headed by the relevant foreign minister: Tariq Aziz on the Iraqi side and Ali Akbar Vellayati for Iran. A UN Security Council report published on 1 August 1988 found evidence of widespread use of chemical weapons by Iraq, and a more limited use by Iran.

On 18 August, the Stockholm International Peace Research Institute announced that the Iran-Iraq war had cost Iraq U.S. $112,000 million, and Iran $91,400 million. And all to no avail....

After the War

Saddam Hussein's capacity for political survival revealed itself, at his cleverest, after the war he had launched but failed to win. As more revelations of his regime's cruelties and other excesses surfaced, so did his initiatives to build up an image of reasonableness and tolerance. Meanwhile, adverse reports were still coming in, from non-Iraqi sources. On 1 August 1988, for instance, the UN published evidence showing that Iraq had recently stepped up its use of chemical weapons (including mustard gas and cyanide) against Iranian troops and civilian targets.

On the 18th, Amnesty International, the human rights organization, accused Hussein's government of deliberately and systematically killing large numbers of Kurdish rebels and civilians. The Kurds themselves claimed that the Iraqi campaign against them had begun on 19 July: the day after Iraq had agreed to a cease-fire in the Gulf War. By late August, indeed, Iraqi forces had spread throughout the Kurdish-held belt of Iraq near the Turkish border. Initially, the Turkish government had attempted to keep Kurdish refugees out, but the Kurds claimed that large numbers of civilians had been subjected to attack by chemical weapons. On 30 August, the Turkish government—to its credit—agreed to accept some 30,000 Kurdish refugees who had gathered on the border.

As part of what was later seen as a propaganda operation, Saddam's son, Udai Saddam Hussein, resigned as president of the Iraq National Olympic Committee and as president of the Iraqi Football Association on 6 November, for "personal reasons." It was reported that after the death of a presidential aide, Kamel Hanna Jajjo, on 18 October, Udai had been banished by his father, either to their home town of Tikrit, or abroad.

On 21 November 1988, as a demonstration of Saddam's impartial sense of justice, even when his own family was involved, he announced on national radio that his son, Udai , until then assumed to be his chosen successor, would be tried for the murder of a presidential aide. As if to drive the message home, he announced less than a week later (on the 27th) an amnesty for political offenders in Iraq and abroad and pledged his government's respect for human rights. He also announced that henceforth the formation of new political parties would be permitted.

Meanwhile, Hussein's new efforts at democratic image making continued. At a meeting of his Revolutionary Command Council (RCC) on 16 January 1989, he announced that a special committee was being set up to draft a new Constitution which would allow the formation of new political parties. He had indeed foreshadowed this move on 27 November the previous year. At a similar meeting on 31 January, it was announced that elections for the National Assembly, originally set for the end of August 1988 , but twice postponed, would be further deferred until 1 April 1989. The announcement delicately hinted at the democratic intentions behind the postponement, which was "to provide more chances for those who wanted to stand."

More amnesties were on the way. The one announced on 30 November1988 was to apply for political offenders inside Iraq and abroad. Against this, on 11 January 1989, an Iran-based Iraqi opposition group claimed that the Iraqi government—Saddam's own— had executed dozens of prisoners who had given themselves up after the November amnesty.

Inside Iraq, hostility towards the regime erupted from time to time. On 8 February 1989, the *Los Angeles Times* reported that Hussein had recently survived an attempted coup, after which a number of senior Army officers were executed. The report was attributed to intelligence and diplomatic sources as well as allegations from Iraqi dissidents. Not surprisingly, the Iraqi deputy prime minister and foreign minister, Tariq Aziz, at a news conference in New York on 9 February, described the allegations as baseless.

19

Lessons of the Second War (1990)

Thwarted in Iran, Saddam still nursed his wounded hubris. Two years after his debacle in Iran, his ambition focused on a much smaller and more vulnerable neighbor: the oil-rich and prosperous emirate of Kuwait, on the northeastern coast of the Arabian Peninsula between Iraq and Saudi Arabia. Iraq, itself a major oil exporter, saw (or claimed to see) Kuwait as a minor but undesirable rival.

On 10 and 11 July 1990, the oil ministers of Iraq, Saudi Arabia, Kuwait, Qatar, and the United Arab Emirates (UAE) met in the Saudi Arabian city of Jeddah. Both Kuwait and the UAE bowed to pressure from their larger neighbors, and agreed to reduce their daily output to a total of 1.5 million barrels each, which meant a cut of 300,000 barrels daily for Kuwait and 400,000 for the UAE.

Outside observers may have seen the agreements as meeting the complaints of major oil states, including Iraq; but Saddam saw them as a bellicose opportunity. Less than a week later, he launched a public attack on the policies of Kuwait and the UAE. During a speech on 17 July, to mark the twenty-second anniversary of Iraq's 17-30 July Revolution, he threatened that "if words fail to protect Iraqis, something must be done to return things to their natural course and return usurped rights to their owners."

He accused some Arab States (which he declined to name) of undermining Arab interests and security on behalf of the U.S. The threat that followed was typical of the man. Iraq, he proclaimed, "will not forget the saying that cutting necks is better than cutting the means of living." The next day, Iraq radio gave details of a memorandum dated 15 July, which, on his instructions, his foreign minister, Tariq Aziz, had sent to Chedli Klibi, the Arab League's secretary-general.

In it, Aziz accused Kuwait of policies aimed at weakening Iraq during the eight-year war with Iran. He claimed that Kuwait had moved into Iraqi territory, setting up oil installations and military establishments. During the war, he asserted, Kuwait had stolen vast quantities of oil from the southern section of the Rumallah oilfield worth, at the time, U.S. $2.4 million.

During that same period, the Gulf States had provided Iraq with interest-free loans, which were still recorded as debts by Kuwait and the UAE. He claimed that these loans were only partly from their treasuries, but in fact came mainly from the profits they made as a result of the drop in Iraqi oil exports during the war.

Had there been a true sense of Arabism during the postwar period, the Gulf States would have cancelled Iraq's debts. Instead, they had glutted the oil market by exceeding their quotas. Since 1987, the Arab States as a whole had lost about U.S.$25,000 million through falling oil prices. Worst hit was Iraq, for having fought a war against Iran, in theory on behalf of all Arab states.

Understandably, Kuwait was alarmed by Saddam Hussein's speech and Tariq Aziz's broadcast. On 18 July, the Kuwaiti Army was placed on alert. A flurry of moves by leaders of the Arab world followed, notably by President Mubarak of Egypt. Ignoring all peace overtures, Saddam Hussein massed an Army of 100,000 on Kuwait's northern border. On 2 August, at 2 A.M. local time, 30,000 of them, including the elite Republican Guards, crossed the Kuwaiti border. Later that day Iraq claimed that its troops had been invited into Kuwait by internal "revolutionaries"—a claim soon shown to be false.

If Saddam Hussein had been allowed to consolidate his rapid victory over Iraq's tiny neighbor, he might have gone down to history as a successful Arab conqueror, if only on a temporary basis. However, the Iraqi adventurer had reckoned without the United Nations and, especially, without the rapid and effective response of the United States under the guidance of President George Bush.

On 3 August—on the morrow of the Iraqi invasion—the U.S. and the United Kingdom announced that naval vessels were being sent to the Gulf. At a meeting in Moscow, the U.S. secretary of state, James Baker and the Soviet foreign minister, Eduard Shevardnadze, jointly condemned the invasion and called for a world ban on arms sales to Iraq.

The crisis faced President Bush of the United States with both a challenge and an opportunity. The challenge seemed clear enough. The U.S. was the West's only superpower but had been defeated, not long before, by a small communist country, Vietnam, admittedly with considerable assistance from the communist world's rival superpower, the Soviet Union. Bush was aware of the assistance the Soviets had been providing for Saddam Hussein, although they were clearly not about to assist Iraq in a war with the West.

The opportunity open to the U.S. was to play a leading role in defeating Iraq and forcing Saddam to pull out of Kuwait. But there was no question of the U.S. single-handedly imposing the West's will on Saddam by invading Iraq. What the U.S. could do, and in fact went on to demonstrate, was to lead an international effort to bring Saddam to heel. Moreover, as Bush rightly judged, this could only be done through the United Nations. For doing this, President Bush deserves genuine credit.

Western action was in fact remarkably fast. On 3 August, the U.S. and the U.K. announced that they would be sending naval vessels to the Gulf. On the 6th, the UN Security Council announced that it had passed Resolution 661 calling for wide-ranging sanctions against Iraq and the small country it had just annexed: Kuwait. Next day, the U.S. Defense Secretary, Dick Cheney, announced that the U.S. had decided to send ground units and combat aircraft to friendly Saudi Arabia, as part of a multinational force.

Despite these clear signs of Western readiness to use force, Iraq formally annexed Kuwait on 8 August. Next day the Soviet Foreign Ministry formally condemned Iraq's annexation of its tiny neighbor but warned against any military action without UN approval. Already, as we have seen, the U.S. had made it clear that no action would be taken without such approval.

The 11th brought an announcement from Washington that the U.S. would enforce a naval blockade against Iraq. Psychologically, the Bush administration had already decided what to do. Next day, James Baker announced that the U.S. had received a formal request from the deposed Kuwaiti government to enforce UN sanctions under the self-defense Article 51 of the UN Charter.

Evidently determined to avoid a U.S. move to take command of any military force, the Soviet Union called for a joint command in the Gulf under the authority of the UN.

On 17 August, Saddam Hussein, clearly minded to rid himself of an earlier crisis before the new one had grown uncontrollably, informed the Iranian president, Hashemi Rafsanjani, that within three days Iraqi troops would begin to withdraw from Iran.

Every day that August brought news relevant to the Iraqi crisis. The most important items were these. On the 21st, the Western European Union met and several of its member countries confirmed that they were sending troops to the Gulf. The most important of them, at that stage, was France, whose president, François Mitterrand, at last announced that French ground forces would be sent to the battle scene.

On the 22nd, for the first time since 1968, President Bush issued an executive order calling up U.S. military reservists. The 30th brought financial news of direct interest to Washington. Bush unveiled an "action plan" under which the richer European, Asian and Gulf powers would share the costs of a prolonged blockade of Iraq, as well as of the military intervention. Unexpectedly, distant but friendly Japan announced that it would contribute U.S.$ 1,000 million to assist the multinational force.

In the event, President Bush's initiative yielded a coalition of 29 military forces, totaling 705,000 armed men. Not surprisingly, the U.S. force, totaling 500,000, outnumbered the twenty-eight other forces by more than two to one; the second largest contribution came from a British contingent of 35,000—less than one-tenth of the American one. France would send only 13,500 men. The offensive, code-named "Operation Desert Storm," began on 16 January 1991. By the end of January, the Allies could claim to have flown more than 30,000 sorties. The favored air weapons were highly accurate cruise missiles and laser-guided bombs. On the Iraqi side, the favored defense weapons were Soviet-made Scud missiles, descendants of the wartime Nazi V-2 rockets.

The main purpose of the allied coalition was not to overthrow the Saddam Hussein regime, but to liberate Kuwait from Iraqi control. This relatively limited liberation was in fact achieved on 26 February, two days after an allied ground offensive. Paradoxically, Saddam Hussein announced on that same day—26 February 1991—an Iraqi "victory" and the withdrawal of Iraqi troops from the oil-rich Emirate where the trouble had begun. All allied military operations were suspended two days later.

As in the other chapters of this book, the military side of war is not the main focus of attention. Inevitably, the questions that predominate after a major war are political. Militarily, this was not a major war; politically, it was. In the defeated country, the predominant questions are: who is in power? Has a new ruler stepped in?

Has the old one stepped down? In Iraq one name answered all three questions: Saddam Hussein. He was still there, still in power. The tyrant whose attempt to take over neighboring Kuwait had failed, as had his attempt to help himself, in Iraq's name, to Kuwait's export-able reserves of oil. Yet challenges to his tyrannical personal rule were not lacking.

In the aftermath of Iraq's military defeat, in late February 1991, a series of guerrilla uprisings broke out, all with the object of remov-ing Hussein from power. The first of them started on 1 March, one day after the suspension of Allied military activity, when crowds of anti-Hussein demonstrators in the southern city of Basra targeted local leaders of the ruling Ba'ath Party and members of the security force. Simultaneously, fighting broke out between rival army units in and around the city. The fighting soon spread to the Shia holy cities of Karbala and Najaf. Also involved in that area was Nasiriyah.

It was soon clear that Saddam Hussein's international defeat was being taken as a signal for the Shia sect of Islam to overthrow and marginalize the Ba'athists. One of the main Shi'ite groups was backed, and indeed dominated, by neighboring Iran. Its polysyllabic name was "the Supreme Assembly of the Islamic Revolution in Iraq" (SAIRI) led by Hojatolislam Seyyed Mohammed Bakr al-Hakim. The SAIRI rebels were calling for the establishment of an Islamic government in Iraq, to succeed Saddam Hussein's under Hakim's leadership.

The SAIRI rebels were not alone, however. Another rebel group had risen in the Kurdish northern provinces. It did not take the Iraqi armed forces long to suppress both uprisings. By the end of March the southern rebels were overwhelmed, while a number of northern towns had been recaptured from Kurdish guerrillas. As the Iraqi army advanced, large numbers of Kurdish rebels fled towards the Turkish border.

To uninvolved or pro-Hussein observers it did not take long to note that a major reason for the speed of the rebellions and also for their rapidly demonstrated failure to oust Saddam Hussein was the unexpectedly passive attitude of the victorious UN forces and espe-cially of their leader, the United States. When military action was suspended on 28 February, the UN forces occupied some 15 per-cent of Iraqi territory. While the Gulf War was on, the United States encouraged the Iraqis to overthrow the tyrant—Saddam Hussein—who had led them into bloody conflict. Now the main concern of the

U.S. was becoming clear: to avoid the disintegration of Iraq after the Gulf War.

This was not the attitude any of the rebel groups had anticipated. They appealed for U.S. aid, on the assumption that it would naturally be forthcoming. The reason for America's non-intervention, dictated from Washington, DC, was that it would be wrong for the Americans to involve themselves in Iraqi affairs (even though no greater involvement than a war could be imagined). This said, the Americans did intervene to relieve the rebels, but only in minor ways. For example, on 22 and 24 March U.S. pilots shot down two Iraqi fighter jets operating over northern Iraq, Their stated reason for so doing was extracted from the interim cease-fire arrangements, which forbade Iraqi use of fixed-wing military aircraft (though not of helicoThe net effect of this (in my view misguided) postwar pacifism was to help keep Saddam Hussein in power, thereby betraying the whole purpose of the Gulf War, as the international public broadly assumed it to be.

The most controversial policy attitude, however, was far more restrictive than a ban on shooting down Iraqi helicopters. A ban on the use of force after victory also extended to the possible removal of the tyrant whose tyranny had publicly justified the U.S. and UN intervention in the first place.

This of course is, or is tacitly held to be, a taboo subject. It is apparently acceptable to mobilize an international force, invade a "rogue" country and bomb its cities from the air, thus causing the deaths of thousands of innocent civilians; but immoral and unacceptable to kill the tyrant who caused the crisis in the first place and had demonstrated his total contempt for human life. Moreover, it was known that Saddam Hussein would not have attacked Iran or invaded Kuwait unless he had been able for years to stockpile Soviet weapons. The Cold War was still on, and nobody was suggesting that the NATO powers should invade the Soviet Union; but since it was demonstrably acceptable internationally for the leader of the West—the United States—to make war on Iraq, why should it be unacceptable to bring Iraq's leader to trial as a war criminal? In other words, why should Saddam's right to life be respected while he had ordered the killing of his opponents?

I raised both these controversial issues in the column I wrote at that time in the influential New York magazine *National Review*:

In late 1973, Saddam personally negotiated the Iraqi-Soviet security and intelligence deals with Gorbachev's mentor, KGB chief Yuri Andropov. In short, Andropov gave Saddam the recipe for absolute power; in return, Saddam gave the Soviets the means for relatively easy spying, notably on Saudi Arabia, through Iraqi embassies in neighboring countries. (Specific details followed.)[2]

...[Do] we "liberate" Kuwait, then sit and watch while Saddam's forces massacre the Kurds? Or do we rid the world of an evil and bloodstained tyrant?[3]

More relevant than the views of a Western columnist and author are those of an American politician and former senior adviser to President Bill Clinton's cabinet: George Stephanopoulos. In a column entitled "Why We Should kill Saddam" in December 1997— nearly seven years after the end of the Gulf War, he wrote:

In the middle of a crisis with Iraq during President Clinton's first term, I wondered aloud in an Oval Office meeting about the prospects of killing Saddam Hussein. Before I could finish the sentence, the then national security adviser, Tony Lake, looked up to the light fixtures and said: "He was just kidding. We're not planning anything like that." Of all the words you just can't say in the modern White House, like "shred this," none is more taboo than assassination.[4]

Stephanopoulos recalled the American efforts to kill Fidel Castro with exploding cigars or poisoned scuba suits, which he characterized as "bordering on the comical." He referred to the Church Committee's revelation of CIA abuses in Cuba, Chile and the Congo. Faced with such publicized failures, President Ford had signed a sweeping one-sentence executive order: "No person employed by or acting on behalf of the United States government shall engage in, or conspire to engage in, assassination."

The White House adviser turned columnist went on to express the view that actions regarded as unlawful and unpopular with the allies of the U.S. were not necessarily immoral. Since he was no longer in the White House, he felt he could now say what he could not in the past: that the assassination option should be considered seriously. Since the Gulf War coalition was now teetering and Saddam's capacity to inflict mass destruction had not been eliminated, killing him could prove to be "the more sensible—and moral—course in the long run."

Certainly, the removal of an evil dictator would have been of enormous and undeniable benefit to the people of Iraq, to their national neighbors and to the Western world. Obviously, though, this could not be a technically simple operation, however desirable a successful attempt would have been. The hypothetical options are worth analyzing.

Of the possible options available, the simplest and cheapest one would probably be the most risky, and liable to fail: to recruit and/or bribe an assassin in the dictator's entourage. A member of that political entourage would probably be unrecruitable, even if driven by outrage and ambition to take over from the man to be removed. No Western diplomat would have the right degree of access, although it is not unthinkable that a skilled member of an Allied secret service (whether of Britain's MI-6, America's CIA or France's SDECE) could be entrusted with the task of recruiting a member of one of Iraq's various services, the main one of which was Al Mukhabarat The recruit might then find an occasion to take advantage of surprise and kill the dictator and whoever else might be within range.

Merely to state the elements of this imaginary plot is to see that it is virtually impossible to realize. An alternative device would be the "poisoned chalice": a beverage or (less likely) a dish, enriched (if that word is permissible) by a deadly poison.

A far more practical plan, especially immediately in the wake of a total military defeat of the enemy, would have been a major commando operation designed to take over the leader's ultimate refuge, seize him and either execute him on the spot or capture him alive with the ultimate aim of bring him to justice. The timing of such an operation would be crucial. It would have to take place to coincide with the total defeat of the enemy. While the possibility of the defeated leader committing suicide cannot be ruled out, neither can the possibility of seizing and removing him for a very public trial, leading to his execution and designated to discourage any other hyper-ambitious tyrant from following in the footsteps of the highest ranking captive.

Although the closest, most intimate members of his bodyguard might remain loyal to the last (his last), they could not, in a final confrontation, save him from his Western enemies.

It is not clear from available material to what extent the two main American leaders involved in the Gulf War examined the alternatives discussed above. It can probably be assumed that the Allied commander-in-chief, General Norman Schwarzkopf, would have given careful consideration to all the available alternatives; but unlikely that the same would be true of General Colin Powell, chairman of the Joint Chiefs of Staff,[5] who was made aware on the spot (that is, in the White House) that the president, George Bush, would veto any such initiatives. If a negative decision of this kind is taken, there is indeed no point in discussing the practical details of the rejected plan.

The price paid for this negative decision is almost immeasurable, and the weight of it fell not on Saddam Hussein or his closest team and family, but on the long-suffering Iraqi people. Technically, Saddam had been defeated, but he remained in power. The UN sanctions hurt the Iraqi people but not the dictator who had provoked them. By mid-1995, inflation had reached 2,400 percent.[6] Street crime had soared and at night the streets of Baghdad were deserted, except for roaming gangs. Saddam's taste for luxury continued unabated. By the late 1990s, the number of his personal palaces was reported to have reached more than two dozen.

Iraq's health services and educational institutions were degraded. Having risen in response to President Bush's appeal, some 50,000 Shi'ites had been massacred "while U.S. troops stood by," according to Khidhir Hamza, an Iraqi nuclear scientist. Meanwhile, Saddam Hussein pushed on with his program for nuclear armaments and ballistic missiles.

In the light of these postwar developments, it is not easy to   offer a historically valid estimate of the value to the West and to civilization of the American and international decision to launch Desert Storm. A mere 148 American soldiers, Marines and airmen were lost, although some 60,000 Iraqi troops were captured. The price, then, was remarkably low. And so were the political gains. To have remained in power was a huge, though militarily negative, victory for the Iraqi tyrant.

Two contradictory lessons emerged from the second Gulf War (1990-1991): how to build an effective United Nations coalition to launch a war with international approval; and how to survive defeat by remaining in power. The hero of the first lesson was President George Bush (Sr.) of the United States; the villain of the second, President Saddam Hussein of Iraq. Sadly, Bush's "heroism" was badly marred by his failure to sanction the removal of the villain.

20

Lessons of Kosovo (1998-)

On 28 June 2001, the former president of Yugoslavia, Slobodan Milosevic, was handed over to the United Nations International Criminal Tribunal in The Hague, capital of the Netherlands. He had been indicted on charges of crimes against humanity in May 1999 and had been in prison in his own country's capital, Belgrade since his arrest on 1 April that year on charges of corruption and abuse of poThe arrest, and especially the handover, made history: it was the first time a former head of state had been arraigned under international justice, and the charges he now faced were unprecedented, relating to his actions in 1999 against ethnic Albanians in the disputed Serb province of Kosovo.

Carla del Ponte, the Swiss-born chief prosecutor at the United Nations War Crimes Tribunals at The Hague, had made known her wish that the indictment would be extended to cover earlier Serb incursions in Bosnia and Croatia.

The indictment included the following charges:

92. The forces of FRY [Federal Republic of Yugoslavia] and Serbia have in a systematic manner, forcibly expelled and internally displaced hundreds of thousands of Kosovo Albanians from their homes across the entire province of Kosovo. (....)

95. Throughout Kosovo, the forces of the FRY and Serbia have harassed, humiliated and degraded Kosovo Albanian civilians through physical and verbal abuse. Policemen,soldiers and military officers have persistently subjected Kosovo Albanians to insults, racial slurs, degrading acts, beatings and other forms of physical mistreatment based on their racial, religious and political identification.

Article 97 accused Milosevic personally (and four of his entourage) of actions resulting in the forced deportation of 740,000 Kosovo civilians. A long list of systematic killings of civilians followed.

Long before its inception on 29 November 1949, the Federal Republic of Yugoslavia had been a cauldron of rival nationalist aspirations. Unity through force had been imposed by the Communist-nationalist leader, Marshal Tito (Josip Broz) who, during World War II, had defeated his pro-Serb rival, Mihailovic—later executed.

There was always a desire for independence among the Yugoslav entities: Croatia and Slovenia broke away from the Federal Republic; and so did Macedonia, followed by Bosnia-Herzegovina. Kosovo, bordered on the southwest by Albania, had a predominantly Albanian, and Islamic, population. In the early 1980s, Albanian rebels exploded bombs in the capital, Pristina, followed by attacks on members of the Serbian and Montenegrin minorities, which at that time together constituted just under 15 percent of the population. One result, not unexpected, was poor relations between Yugoslavia and Albania. The former accused Albania of interference in Kosovo, and the latter of unjustified threats to its country.

On 28 September 1984, a Yugoslav court sentenced a twenty-five-year-old student, Chanan Memeti, to six years in jail for spying for Albania. In February 1985, another student of Albanian origin was jailed for twelve years on charges of spying for an unnamed country.

Over the next two years, there was an increasing flow of emigration by Serbs and Montenegrins from Kosovo, and the central Yugoslav government introduced measures designed to speed the pace of ethnic integration of the Albanian population. These measures included compulsory bilingual education, employment and housing incentives to encourage recent emigrants to come back; and stringent penalties on Serbs and Montenegrins trying to sell their homes to ethnic Albanians.

A growing spiral of violence developed in January 1998, involving the Serb police and the clandestine Kosovo Liberation Army (UCK) operating in the mainly ethnic Albanian province. On the 4th, the UCK issued a statement describing itself as the "armed forces of the Kosovo Albanians." It reiterated its determination to fight for the unification of Kosovo with Albania.

Three protests were made, condemning the Serbian police for using violence against peaceful student demonstrators. The protests came from the Serbian Orthodox Church, the Albanian government and the Organization for Security and Cooperation in Europe (OSCE). The students had in fact been protesting against the closure of Albanian language schools in Pristina and two other towns on 30 December 1997. At least thirty-two students were injured in clashes with the police. It emerged later that four policemen had been shot dead in the village of Likosane in central Kosovo. In a retaliatory action on the 28 January and 1 March 1998, twenty-four ethnic Albanians were killed. Their compatriots in the villages claimed that the dead were all civilians. They alleged that eleven of the victims had been dragged from their home, beaten, and shot in cold blood.

In Strasbourg on the 28 January, the Council of Europe's Parliamentary Assembly passed a resolution condemning "Serbian repression of the ethnic Albanian population of Kosovo." The resolution added that Yugoslavia's policies had led to armed resistance in the region. On 2 March, some 50,000 ethnic Albanians joined in a mass demonstration in Pristina. Next day, tens of thousands of Albanians gathered in the capital to bury their compatriots killed in the villages.

The United States and the European Union issued strongly worded denunciations of the Serbian action, while criticizing the UCK's separatist aspirations and violent methods. Not surprisingly, Russia declared that the threat of force or further sanctions against Yugoslavia was unacceptable. In early March China vetoed the issuing of a UN Security Council statement on Kosovo, objecting to UN interference in Yugoslavia's internal affairs.

The unrest continued in April. Indeed, the clashes between ethnic Albanians and the Serbian army and paramilitary units intensified. The leading Western powers—the U.S., Britain, France, and Germany, together with Russia, had called on the Yugoslav government to withdraw special Interior Ministry police from Kosovo, but President Milosevic defied the request and turned down demands for him to open internationally mediated talks with Kosovo separatists.

Throughout April, thousands of ethnic Albanians staged peaceful demonstrations against Serbian rule. On the 9th, more than 30,000 ethnic Albanians were on the march.

Already, on 31 March, the UN Security Council had adopted Resolution 1160, which imposed an arms embargo against the Yugoslav Republic. The same Resolution urged the authorities in Belgrade and the leaders of the Kosovar Albanian community "urgently to enter without preconditions into a meaningful dialogue on political status issues."

After a meeting on 29 April, all members of the contact group, with one exception, agreed to freeze the financial assets of the Yugoslav Republic. The exception, of course, was Russia. The group also warned the Republic that it would cut off all public investments in the Republic if the government failed to enter into meaningful negotiations with the ethnic Albanians by 9 May.

The ethnic Albanians, however, were not in a negotiating mood and refused, on 9 May, to enter into bilateral talks with the Serbian authorities. Already on 7 April, they had rejected an offer of dialogue with Serbia's president, Milan Milutinovic, reiterating their previous refusal to join any such discussion unless and until the special police forces had been withdrawn.

On 27 April the EU foreign ministers had confirmed their adherence to the UN arms embargo against the Yugoslav Republic and had decided to ban the export of police equipment and armored vehicles to Serbia. Moreover, any requests for visas from Serb officials would be banned. Five days earlier, the UN Human Rights Commission had condemned Serbian officials for violently repressing the expression of political views in Kosovo. The Resolution was approved by forty-one members of the fifty-three-country panel; the remaining twelve, including Russia, had abstained. It called for the withdrawal of special Serb police from Kosovo, and for Yugoslav officials to arrest indicted war criminals.

On 23 April, Serbia had held a referendum on whether the people of their country approved of foreign involvement in talks on the future of Kosovo. The referendum was, not surprisingly, boycotted by ethnic Albanians in Kosovo. An official statement on the 24th announced that 94.73 percent had voted against foreign intervention. Next day, the opposition Belgrade Centre for Free Elections and Democracy challenged the official figure of a 73 percent turnout. A spokesman for the group drew attention to the fact that ethnic Serbs constituted only 63 percent of Serbia's population, and that the leaders of most ethnic organizations had called for a boycott of the vote.

The Montenegrin president, Milo Djukanovic, had said on 13 April that the referendum, proposed by President Milosevic, was a form of "collective suicide." He had also called on the international community to support his "efforts to form a bloc of reformist forces in Yugoslavia as a whole to bar the way to Milosevic.

On 31 March 1998 the UN Security Council imposed an arms embargo against the FRY; and on17 April, the Albanian People's Assembly appealed to NATO to deploy troops on its territory to avoid the conflict "spilling over."

Responding to the appeal, NATO announced on 28 May that it would draw up contingency plans for the deployment of troops along the Albanian border with Kosovo, to prevent the spreading of the conflict. At this stage, however, NATO's plans did not envisage military intervention. The deployment of troops along the border was aimed at stopping the smuggling of weapons and a guarantee of Albania's territorial integrity, plus deterrent large-scale military exercises along the border.

Throughout July, the Serbian security forces launched a series of sustained attacks against ethnic Albanian rebels. As it happened, this sweeping offensive (as it came to) turned into one of the most successful of the government's campaigns. By the end of the month, Milosevic called a halt on all military action. On the 30th, Milosevic held talks with a three-member party of EU envoys. Milosevic declared, repeatedly, that he was halting all military action and was willing to discuss autonomy for Kosovo, but not independence.

On the 6th, diplomatic observers from nine countries (including the U.S., the U.K. and Russia) had begun patrols in Kosovo. On the 28th, the U.S. ambassador to Macedonia, Christopher Hill, was reported to have reached a deal with Kosovan representatives on the make-up of a Kosovan team prepared to negotiate with the Serbian authorities. On the 15th, Ibrahim Rugova's Democratic Alliance of Kosovo (LDK) attempted to convene a legislature of the "Republic of Kosovo" but his attempt was broken up by Serbian police.

Fighting resumed on a fiercer level throughout August. The result was a near rout of the Kosovo Liberation Army. Although Milosevic had declared an end to the offensive, at least six ethnic Albanian villages had been destroyed by the Serb forces in early August. On the 24th, the UN Security Council called for an immediate ceasefire in Kosovo; but the plea was ignored by the Serb forces, which continued their ruthless attacks on Albanian ethnic villages.

A five-day NATO operation code-named Cooperative Assembly 1998, involving 1,700 NATO troops from fourteen countries was launched in early August; but the fighting went on. Media reports of mass graves of massacred ethnic Albanians were investigated, but EU observers found no evidence to support the stories.

Nevertheless, tensions continued, and it was clear that President Milosevic had decided to strip Kosovo of its autonomy. The ethnic Albanians staged protests, and Milosevic sent in troops to restore order. Kosovo, however, was not the only province in rebellious mood. Slovenia and Croatia declared independence and the United States warned Milosevic against any further violence against Kosovo Albanians.[1]

Nor was Kosovo the only scene of violent nationalism. Bosnia-Herzegovina broke away, and Croats and Serbs engaged in "ethnic cleansing." Faced with this worsening crisis, the United Nations sent an armed Protection Force, known as UNPROFOR, charged with protecting minorities, provide food for hungry civilians and protecting "safe" areas.

NATO too intervened once more, by sending a force of 60,000, consisting mainly of Americans, British and French, to restore and maintain order. NATO air strikes began on 24 March 1999. These attacks were by no means always successful. Despite the technological progress, air bombardment remained a dangerously inaccurate weapon, as two incidents in particular demonstrated in May 1999. As it happened, both took place on 7 May, as follows:

1. A cluster bomb dropped on an airfield at Nis, in southeast Serbia, hit a civilian area instead of its target. A hospital and an outdoor market were hit. Fifteen people were killed and seventy injured.

2. A more sensationally regrettable incident happened that evening, when U.S. Stealth bombers from the Whiteman Air Force base in Missouri NATO bombers mistakenly attacked the Chinese embassy in Belgrade. Four Chinese nationals were killed, and twenty others injured. The four dead included the Belgrade correspondent of the Guangming daily, his wife, and a reporter for the official Chinese Xinhua news agency.

The U.S. authorities immediately admitted that the attack had been made in error. The NATO spokesman, Jamie Shea, declared that the Alliance deeply regretted the loss of life and injuries and blamed the bombing on an error in "the intelligence process." Next day, in Washington, a joint statement was issued by the U.S. Defense Secretary, William Cohen, and the director of the Central Intelligence Agency (CIA), George Tenet. The attack, they declared, "was the result of neither pilot nor mechanical error. Clearly, faulty information led to a mistake in the initial targeting of this facility."

U.S. officials later claimed that the basis of the error was an outdated map. The pilots reportedly believed that their real target was the Yugoslav Federal Directorate of Supply and Procurement.

Not surprisingly, the attack was greeted with anger in China. There were officially encouraged mass protests. Crowds gathered before the U.S. embassy, which was damaged (and the United Kingdom one as well, since both nations were part of NATO) when the demonstrators hurled bricks at the buildings. Finally, on 30 July, the U.S. government agreed to pay $4.5 million to the families of the victims. As if to negate the apology, the U.S. government claimed that the "humanitarian" payment was entirely voluntary and did not represent an acknowledgment of legal liability. Despite these unfortunate events, U.S. bombings on NATO'S behalf continued. Among them was an incident on 20 May when a stray NATO bomb killed or injured patients and staff in a Belgrade hospital, also damaging the Swedish, Norwegian, and Spanish embassies. Collectively, these incidents demonstrated, once again, the failings of air attacks, even more than fifty years after World War II.

More importantly still, the whole episode of NATO intervention in Yugoslavia raised a major, if officially unstated, question: was NATO's intervention legitimate in the light of its own mandate? A few points in NATO's complex history need to be recalled. The most important, of course, relates to a major problem that no longer exists in its original form: the threat of an invasion of the democratic countries of Western Europe by the Soviet Union. To that end, the North Atlantic Treaty was signed on 4 April 1949.

Curiously, in the light of the situation then prevailing, the Soviet threat was not mentioned specifically. As Article 5 of the Treaty put it: "The parties agree that an armed attack against one or more of them in Europe or North America, shall be considered an attack against them all...."Any such attack would be met, if necessary, with armed force, and all measures would immediately be reported to the United Nations Security Council. The collapse of the Communist regime in the Soviet Union at the end of 1991 removed the original (though unstated) raison d'être of NATO, while leaving its defensive potential against any future military threat untouched. Not surprisingly, the original NATO Treaty made no provision for intervention in Yugoslavia, or indeed in other Balkan countries.

Clearly, then, NATO acted beyond its mandate by intervening with military force in Yugoslavia. It could also be argued that it did so in Afghanistan in response to President George Bush's appeal for solidarity after the Afghan-Islamic terror attacks or 11 September against the World Trade Center buildings in New York. And yet, in both cases, NATO's intervention was necessary in defense of Western values.

There is therefore an urgent need for a reassessment of NATO's role in the new world situation resulting from the collapse of the Soviet Union and the spectacular expansion of international Islamic terrorism. NATO is still needed, but not NATO as it was created and assembled in 1949.[2]

Part 6

Looming Threats

21

The Chinese Threat (Part I)

Post-Mao China became the world's ultimate political paradox: a Communist regime with a capitalist economy. In the last years of his dictatorial life, Mao Zedong denounced his fellow-veteran of the Long March, Deng Xiaoping, as a "capitalist roader" and, to make the point clear, had him frog-marched in public by his youthful Red Guards, with a donkey's bonnet on his head.

This unpleasant spectacle was inflicted on the Chinese people and the outside world during the "Great Proletarian Cultural Revolution" of the 1960s. To be fair, Mao had a temporary attack of comradely remorse in 1973, when he ordered the ruling Communist Party to rehabilitate his old comrade of the Long March; but three years later, he reverted to his natural disloyalty to a comrade and ideological loyalty to his own ideas by denouncing Deng as a "capitalist roader." Three years on he died, on 9 September 1976, thus relieving the party of its painful ideological dilemma.

Eleven years later, in 1987, Deng proved Mao right by launching his own sensationally successful capitalist experiment. In the ensuing years, China emerged from its long era of economic stagnation, reaching 13 percent economic growth in 1992. The built-in paradox was that the anti-capitalist doctrine inherited by Mao from his Soviet tutors was left untouched, although henceforth violated on all sides by the rising capitalist magnates and investors.

It is of course impossible to grasp the history of Communist China without looking back at its origins and at the determining role of Mao Zedong. Mao has two rivals only in a choice of tyrannical monsters in the twentieth century: Stalin and Hitler. Two rivals whom he leaves far behind, as the figures show. In the Soviet Union, the total number of deaths blamed on the regime was 20 million. In Communist China, the almost incredible total was 65 million. These stupefying totals leave Hitler's 7 million a long way behind.[1] I quote these figures simply as an illustration of the ruthless tyranny inflicted on his people by the Chinese Communist leader.

Born in 1893, in a village in Hunan, Mao joined the Nationalist revolutionary Army, led by Sun Yatsen, in October 1911, aged less than eighteen. A year later, he left and resumed his studies. Having graduated, aged twenty-five, he went to Peking (now Beijing), and worked in the University Library, where he met two determined Marxists: Li Tachao and Chen Tuhsiu. The latter, in particular, converted Mao to Marxism when they met again in Shanghai in April 1920. The three young men, and ten others, at a secret meeting in the same city in July 1921, jointly founded the Chinese Communist Party.

Confusingly, the new CP agreed to cooperate with the Kuomintang (Nationalists) the following year, but Sun Yatsen died in Peking in March 1925 and two years on, under the leadership of Chiang Kaishek, the Nationalists started a purge of Communists as Chiang formed a government in April 1927.[1]

The split became definitive on 1 October that year, when Mao, in Peking, proclaimed the People's Republic of China. A few weeks later, he flew to Moscow to negotiate a Sino-Soviet Friendship Treaty.[2]

In the two ensuing decades China was the extended scene of half a dozen competing wars: a Sino-Japanese war (in which, at times, the Nationalists and Communists collaborated); a civil war between the Communists and Nationalists; two invasions: Russian and Japanese. Although Mao emerged the victor in his major civil war clash with Chiang, his relations with Moscow and within his own Party varied from time to time. In June 1932, for instance, Mao was forced into temporary retirement. In May 1934 the Communist Party started preparations to abandon their base in Kiangsi without bothering to consult Mao, although he had been re-elected chairman of the government in January.

In October of that year, the Communists broke through the Nationalist blockade and the famous Long March began. It ended in October of the following year. The "marchers" had covered 6,000 miles, primarily to relocate the Communist revolutionary base from southeast China to northwest China. Throughout their journey, the Communists had repulsed Nationalist attacks: crossing twenty-four rivers and eighteen mountain ranges to reach the northwestern province of Shensi.

One of the highlights of this complex period was the Japanese occupation of Manchuria in 1931-32. Another was Chiang Kaishek's control of Kwangtung province on 19 July 1936. Over the next nine years, the Japanese advance continued, marked by the seizure of Peking in July 1937 and Tientsin a day later (on 29 July).

The Japanese went on to take Shanghai in November 1937 and Nanking in December. After several further victories, the Japanese invaders occupied Canton on 21 October 1938 and Hankow four days later. When World War II broke out in Europe on 3 September 1939, Japan seized its opportunity to resume its undeclared war in China. On 30 March 1940, the Japanese set up a puppet Chinese government.

With many other things on their minds in October 1942, the U.S. and Britain announced their relinquishment of extra-territorial rights in China. Slightly less than a year later—on 13 September 1943— Chiang Kaishek was named president of the Chinese Republic, while retaining his Nationalist title as commander-in-chief of the Chinese Army. The reality, however, was that China remained divided between Chiang's Nationalist forces and Mao Zedong's Red Army. Japan's sudden collapse in August 1945, however, enabled the Red Army to overrun most of the Northern provinces.

Unexpectedly, the Nationalist premier, T. V. Soong, concluded a Treaty of Friendship and Alliance with the Soviet Union on 14 August 1945. A fortnight later, negotiations began between Chiang and Mao, but by 11 October they had collapsed, and before the end of that month heavy fighting had broken out between the Nationalists and Communists in North China.

In mid-December, the United States sent General George Marshall to mediate a truce between the rival Chinese governments. But when the Nationalists turned down a Communist request to share control over Manchuria with them, all-out civil war broke out again, on 14 April 1946.

On 10 October, Chiang Kaishek was re-elected president of China —not by the electorate, but by the Kuomintang. The civil war went on, with successes on both sides. On 19 March 1947, the Nationalists captured the Communist capital of Yenan province. But by the end of 1947, the Communists had gained complete control over Manchuria. Thereafter, the Nationalists headed rapidly towards collapse. By the end of 1948, most of northern China was under Communist control, and the Nationalists had become seriously dependent on American aid. Early in 1948, the generous U.S. allocated an extra $400 million to the Kuomintang.

Beijing fell on 21 January 1949, and that day Chiang Kaishek resigned as president. On 20 April, Mao Zedong called for the formation of a Communist-Nationalist coalition cabinet—but of course without Chiang, who was denounced as a "war criminal."

On 1 October that same year, the Communists proclaimed the People's Republic of China in Peking, with Mao as chairman of the administrative council and Chou Enlai as prime minister and foreign minister.

At this stage, it is useful to devote some space to relations between the Communist parties of China and the Soviet Union. In the early stages of China's civil war—in 1927-1928—Stalin had virtually ignored the occasional plight of Mao and his Communist followers.

He changed his mind, however, in a decisive manner, after his "last minute" decision to take advantage of the U.S. atom bombing of Hiroshima on 6 August 1945. Two days later, the Soviet Union declared war on Japan, and started it by invading Manchuria and in effect ordering a large-scale looting of capital equipment. In the late spring of 1946, he also ordered a handover of vast quantities of Japanese arms and equipment to Mao Zedong's Army.

The scale of the handover was prodigious: 1,226 cannon, 369 tanks, 300,000 rifles, 4,836 machine-guns and 2,300 motor vehicles.*

Stalin also arranged the transfer of 100,000 trained North Korean troops into the forces of Mao's military commander, General (later Marshal) Lin Piao. After the Allied victory over Japan ("Allied" meaning the wartime alliance between the West and the Soviet Union), Stalin invited Mao to Moscow, to join in the celebrations of his birthday on 21 December. Not surprisingly, Mao accepted the invitation. Unexpectedly, however, he spent nearly nine weeks in Moscow, mainly, it appears, because Stalin kept him waiting for the two long conversations they had, on 16 December 1949 and 22 January 1950.

---

*These figures, originally given by the Kuomintang, were officially confirmed by Beijing in 1957. See Brian Crozier, *The Man Who Lost China* (op.cit.) p.291. I have also cited them in ch. 14 ("China: The Temporary Satellite") of *The Rise and Fall of the Soviet Empire* (op. cit.), p.142.

To Mao's apparent surprise, the Soviet dictator proved remarkably friendly and generous. On the first occasion, he responded immediately and positively to his Chinese visitor's request for a Soviet credit of $300 million. He demurred, however, when Mao raised the possibility of an assault on Formosa: "What is most important here," he said, "is not to give the Americans a pretext to intervene." The second meeting, in the presence of various Soviet and Chinese notables, Stalin agreed to Mao's request for an interest of only 1 percent on the credit offered.[3]

On 14 February 1950, Mao and Stalin signed a thirty-year Treaty of Friendship, Alliance, and Mutual Assistance. The Soviet Union assured China of Soviet support against attack by Japan or any other State cooperating with Japan. And it transferred Soviet interests in the former Chinese railway linking Manchuria and China, without compensation, as well as in Port Arthur and Dairen, "immediately on the conclusion of the peace treaty with Japan, but not later than the end of 1952.[4]

In October 1950, Mao manifested his "loyalty" to Stalin by sending Chinese troops to intervene in the Korean war.*

In February 1957, in one of the most evil of his decisions, he launched his "Hundred Flowers" policy. Under the slogan "Let one hundred flowers bloom, let a hundred schools contend," all Chinese people (meaning, in particular, the intellectuals) were encouraged to speak their minds and their thoughts. Not surprisingly, many did so, after initial hesitations. Having expressed what was in their minds they were persecuted for so doing, from June on: intellectual and political "freedom" had lasted four months.

In May 1958, Mao struck again: launching his Great Leap Forward Movement, intended to bring "People's" China, in one bound, to the top of the world in industrial production. Predictably, it did the opposite, raising the death rate in the countryside from 12.5 percent per thousand in 1958 to nearly 29 percent in 1960. The famine went on into 1961, and by the end of that year more than 20 million died as a result.[5] In December that year, the "Great Helmsman," as he liked to be known, stepped down as State Chairman. The following year, at a meeting of the party's Central Committee, Marshal Peng

---

*He also ordered the For further details, see Part VI, Ch. 3. launching of an intensive anti-American movement throughout China. In September 1954 he convened the first National People's Congress in Peking, which approved a Constitution, with himself as chairman of the sate.

Tehuai attacked Mao's Great Leap Forward policy and was sacked. His successor. Lin Piao, obliged the Great Helmsman by intensifying the cult of the leader, notably by issuing Mao's Little Red Book of his slogans in millions of copies.

In May 1966, Mao launched the Great Proletarian Cultural Revolution, which again produced disastrous results, and the following month, he mobilized the youthful Red Guards for "great destruction." They obliged by beating or killing citizens throughout the vast country. It was during this countrywide rampage that Deng Xiaoping was frog-marched with a donkey hat.

And so on, and on. In 1973, however, Mao restored Deng to his former political position, but not for life. On 8 January, Chou Enlai died, and Mao stayed away from the funeral. In April, there were apparently spontaneous demonstrations in favor of Chou. In another dictatorial quirk, Mao again dismissed Deng Xiaoping. And then, on 9 September, Mao at last provided a real service to his people: by dying.

By this time, Deng was thoroughly disillusioned over the repeated economic and social failures of "The Great Helmsman"; but not, unfortunately, over the absolute "right" of the ruling Communist Party to govern the Chinese People's Republic

In April 1976, demonstrations in favor of Chou Enlai (and by inference against Mao) took place in the vast Tienanmen Square in Peking. The demonstrators were also anti-Deng, for being a "capitalist roader." The demonstrations were repeated in April 1978. By then, Deng was back in power, and supported brutal suppression of the demonstrators.

In contrast to Mao Zedong, Deng Xiaoping took a strong interest in national reunification, which primarily referred to Taiwan, Hong Kong and Macao. He was, of course, well aware of the descending order of resistance on the part of these coveted territories. Taiwan had made spectacular economic progress since Chiang Kaishek's decision to take refuge there, and apart from its considerable armed forces, it enjoyed, de facto, American protection. Hong Kong had been ceded "in perpetuity" to the United Kingdom under the Treaty of Nanking in 1842 but there was no sign that Britain would choose to defend it militarily. The same was true of Portugal's Macao. In the event, after years of negotiations, the two sides reached agreement on the future of Hong Kong on 27 May 1985, under the agreed formula of "one country, two systems." In 1981, Deng had told a group of visiting U.S. senators that, even after reunification, Taiwan would be able to retain its existing social and economic systems, and even its armed forces.[6]

Despite these assurances, the Taiwanese—with American backing—opted for continued independence, under their chosen title, the Republic of China. Its prosperity was a model, not only for giant China but for many other countries. It had also established its democratic credentials. In the general elections of 1 December 2001, the late Chiang Kaishek's Kuomintang (Nationalist) Party had lost its majority in the Legislative Yuan (or Parliament) to the new president Chen Shuibian's Democratic Progressive Party. In sum, there was virtually no substantial desire for reunification with the ancient but still despotic motherland.

22

The Chinese Threat (Part II)

In mid-1949, utterly defeated by Mao Zedong's Communists, the Chinese Nationalist leader Chiang Kaishek withdrew from active politics, though without resigning office. He took his son Chingkuo with him into "retirement" at Chi Kou, his native village.[1] For many years—on and off since 1921—Generalissimo Chiang had been at war of one kind or another: against the Communists, against their Russian supporters, against the Japanese, against rival warlords.

Having lost control over the gigantic mainland, there was only one option open, other than surrender (which he could not contemplate): to seek refuge on the island of Taiwan (or Formosa as it was generally known at that time). For fifty years, the island had been under Japanese control, but the inhabitants still spoke their form of Chinese and read the classical Chinese characters. It was a natural retirement place for the defeated Generalissimo.

From his village retreat, Chiang instructed the Governor of the Central Bank of China, O.K. Yui, to transfer the entire gold reserve of 500,000 ounces to Taipei.[2] By that time, Chiang no longer had the constitutional authority to give any such order, but he was obeyed, presumably because he invoked his authority as director-general of the Kuomintang: Chiang's Nationalist Party. He is thought to have believed that the transfer would enable him to pursue his policy of continued resistance to the Communists. Moreover, in so doing he was depriving his rival and successor, Li Tsungjen, of the financial means to continue resistance to the Communist victors.

But Chiang had not stopped there. In mid-1949, he had begun to transfer some 300,000 of his loyal troops to Taiwan. The withdrawal was completed by 6 December. On 1 March1950, Chiang Kaishek resumed his presidency: not of course over the Chinese mainland, but over the former Japanese colony of Formosa, henceforth to be presented to the world as the Republic of China. Indeed, Chiang Kaishek never repudiated his claim that his government was the legitimate one not only of Formosa but over mainland China as well.

Initially, this oft-repeated claim was an irritant to Formosa's one international defender: the United States. On 3 February 1953, President Eisenhower relieved the U.S. 7th Fleet of its duties of "neutralization" around Taiwan (as we shall call it henceforth). Not surprisingly, however, when it became clear that Communist China would try intimidatory moves against the Nationalist government, the U.S. resumed its protective role.

The first of the intimidatory moves came in September 1954, when the Nationalist government reported heavy artillery fire from the Chinese Communists against the offshore islands of Quemoy and Little Quemoy, which remained under Nationalist control. On 2 December, Chiang Kaishek's government signed a mutual defense pact with the U.S. At this stage, the U.S. commitment was not unconditional. It was confined to Taiwan itself and the Pescadores, but significantly omitted the Taiwan-controlled offshore islands.

On 30 September 1958, the American secretary of state, John Foster Dulles, responding to adverse American public opinion regarding U.S. policy towards China, declared that the United States had no commitment to defend Quemoy and Matsu. Whether or not in response to this declaration, the Chinese Communists gradually reduced their bombardment of the offshore islands over the ensuing months.

Two years earlier, in December 1956, my then employer, the *Economist*, had sent me on an exploratory journey to Taiwan—in retrospect, one of the most productive professional trips of my career. I found well-filled shops, a high standard of literacy and of living and, not least, a calm and apparently happy population the numbers of which had gone up from 7,500,000 to 10,000,000 since the influx of Nationalists from the mainland.

The most important discovery I made, however, was the great land reform pushed through between 1951 and 1953. Astonishingly, three-quarters of the peasants now owned the land they were tilling. The proportion of farmers owning all or part of their land had increased from 57 to 75 percent, while the proportion of tenants had fallen from 36 to 19 percent.

Whereas on the Chinese mainland and in North Vietnam the Communists had murdered landlords on a sickeningly vast scale, the

Chinese Nationalists, in effect, had turned them into industrial capitalists. The government had handed the original landlords 70 percent in land bonds and 30 percent in the form of stocks in government-owned industries.[3]

To give credit where it is due, the land reform program was supervised by the very able vice president of the "Republic of China," General Ch'en Ch 'eng, whom I interviewed on my visit to Taiwan, along with President Chiang Kaishek himself.

From 21 to 23 October 1958, Secretary of State Allen Dulles and President Chiang conferred in Taipei. A joint communiqué stated that the Nationalists would not attempt to return to the mainland by force. A few days later (on 31 October), however, the U.S. State Department revealed that Chiang Kaishek had reserved the right to use force, in the event of a "large-scale" anti-Communist revolt on the mainland. This was of course merely a morale-building statement on the Generalissimo's part. A year later, on 7 October 1959, Undersecretary of State Douglas C. Dillon declared that in the event of a Chinese Communist attack on Taiwan and the offshore islands, they would be risking "total" world war.

23

The North Korean Threat

North Korea's long-time leader, Kim Il Sung, was very much one of Stalin's protégés. His original name was Kim Sung Ju, but the name he chose, in his late teens, was that of "a legendary Korean hero in the Robin Hood mould."[1] When he was seven, his father was jailed for anti-Japanese activity: at that time, Korea was a Japanese colony. At eighteen, he joined the Chinese Communist Party in Manchuria. More importantly, he went to Moscow in 1941 for political and military training, returning four years later as a major in the Soviet Red Army. In 1946—by that time chairman of the Soviet-sponsored Provisional People's Committee—he took over control of the Soviet zone. By that time, Japanese colonizers had been driven out of Korea.[2]

By 1950, the former Japanese colony had been turned into two countries: a Communist one in the north and a possible Western-style one in the south, under the leadership of Dr. Syngman Rhee, a U.S.-educated academic with an American wife.

On 10 January 1950, the Soviet ambassador in the North Korean capital, Pyongyang, T.F. Shrykov, reported to Stalin about a conversation he had had with Kim Il Sung during a reception at the Chinese embassy. The gist of it was that Kim wanted to visit Stalin and ask his permission to "liberate South Korea." Stalin immediately told the ambassador that he was ready to receive Kim. The meeting was held in Moscow in March, and Stalin not only approved of Kim Il Sung's plan but also pledged full support with arms and equipment to meet the North Korean leader's plans.[3]

Having obtained Stalin's backing. Kim Il Sung launched his invasion of South Korea in total surprise. He had chosen a Sunday morning, at 4 A.M. local time, taking advantage of the knowledge that the "main enemy"—the United States—would be totally unprepared for a crisis of this nature. Being a trained Communist, moreover, he provided a perfect example of Soviet-style disinformation. Radio Pyongyang was the natural outlet for an announcement that North Korea had declared war on South Korea because of an invasion of the North launched by "the bandit traitor Syngman Rhee" who, it went on, would be arrested and executed.[4]

Two hours later, Kim rounded off the disinformation exercise in a broadcast claiming that because North Korean territory had been attacked that morning, by South Korean forces, South Korea would have to take the consequences of North Korea's counter-action.

In Washington that sleepy Saturday morning, telephone connections to Korea were closed. Radio messages took over, and within hours the U.S. government and the United Nations were alerted to the major and totally unexpected crisis that faced them. Unable to telephone Washington, the U.S. ambassador in Seoul, John Muccio, cabled the news to the State Department. At the receiving end was Dean Rusk, the assistant Secretary of State, who immediately rang his superior, Dean Acheson; who in turn rang President Harry Truman at his home in Independence, Missouri.[5]

The immediate reaction was to alert the United Nations, which convened its Security Council. Exceptionally, in this exceptional case, the Council met next day—Sunday. As it happened, at this crucial meeting, conditions happened to be ideal for a Western reaction, in that the Soviet representative, Jacob Malik, had decided to boycott the Security Council because China was represented by a Nationalist instead of a Communist.

Because of this short-sighted error, the Security Council was able to escape a Soviet veto, and voted in favor of military action by nine votes to none, with one abstention: Yugoslavia, which had broken away from Soviet domination.

At a second meeting, the Security Council (still without a Soviet representative) passed a further resolution, calling on members to give South Korea whatever help it needed to repel the invasion. The practical outcome was a decision by the U.S., Britain, Canada, and Australia to provide military units to fight on the side of Syngman Rhee's forces.

By 20 June, the South Koreans had been forced to retreat over the Han River south of the capital, Seoul, by now in Communist hands. At this point, General Douglas MacArthur, who had visited the front on President Truman's orders, was back in Tokyo, where he described the Korean retreat as a rout.

The great general was about to live up to his victorious reputation. In the face of criticism and misgivings from most of his military and naval colleagues, he decided on a surprise landing at Inchon,
a port on the coast southwest of Seoul. Despite an approach studded
with shoals and rocks, the landing was a great success, taking the
North Koreans and their Soviet advisers completely by surprise. The
North Korean army virtually collapsed, and by 19 October,
MacArthur's forces had crossed the 328th parallel dividing line and
captured the North Korean capital, Pyongyang.

Immersed in his own glory, MacArthur appears to have dismissed
the threat of Chinese intervention. It duly came, however, between
25 October and 7 November, and broke the general's offensive. U.S.
Military Intelligence—the G-2—turned out to have been deeply mis-
led: it had estimated the number of Chinese forces in Korea at about
60,000, whereas it turned out that the real number was 180,000—
three times as high.[5] In the end, MacArthur's Eighth Army was driven
into full retreat.

After the war, Korea remained divided into two rival autonomous
states: the Democratic People's Republic of Korea in the Communist North and the Republic of Korea in the democratic South, divided from each other by the 38th parallel. In other words, North
Korea remained an appendage to the Communist empire, under
Chinese rather than Soviet control; while South Korea qualified, as
time moved on, to be regarded as an extension of the non-Communist democratic world.

The Communist North, dominated by the dictator Kim Il Sung,
was one of the most ruthlessly brutal of Communist states, with its
gulag and its leader, known under his own chosen title of "the great
leader." His son, Kim Jong Il, who would succeed him after his death,
was dubbed "the beloved leader." and lived a life of luxury and self-
indulgence. In contrast, South Korea, although slow to practice democracy as understood in the West, achieved a remarkably effective
and competitive capitalist system, becoming a highly successful
exporter, notably of high-class cars.

On 13 June 2000 a summit meeting, dubbed "historic" by the
media, took place in the North Korean capital Pyongyang, between
the presidents of the rival States: South Korea's Kim Dae Jung, and
the North Korean de facto head of state Kim Jong Il, chairman of the
North Korean National Defense Council. It was the first of its kind
since the two rival regimes had emerged as independent States in

1948. Foreign observers were impressed by the warmth of the reception organized for Kim Dae Jung, whose motorcade was greeted by some 600,000 cheering North Koreans.

The meeting had been marked by genuinely useful steps, including the reunion of divided families, the repatriation of political prisoners and continued dialogue between the rival States. Further talks took place from 27 to 30 June, but these were non-political, between Red Cross officials, from North and South, at the North Korean resort of Mount Keumgang. The outcome was an agreement allowing relations between families split by the Korean War. As if to confirm the moves towards better relations between the warring regimes, the U.S. on 19 June announced an easing of economic sanctions against North Korea. Already, in September 1999, President Clinton had announced his government's intention to lift economic sanctions after the North had agreed to suspend further tests of ballistic missiles. On 21 June 2000, North Korea confirmed its moratorium on test flights of the said missiles.

The outside world naturally welcomed these developments, but future events would cast a shadow over them. For instance, at the conclusion of a visit to Moscow by Kim Jong Il on 8 August 2001, a joint declaration by him and his Russian host, President Vladimir Putin, called for the speedy withdrawal of U.S. armed forces from South Korea—a reversal of Kim's apparent position at his meeting with South Korea's Kim Dae Jung in June 2000. (It would be reasonable to suppose, however, that the inclusion of this clause in Moscow would have been at Putin's request.

On 19 November 2001, South Korea's defense minister, Kim Dong Shin, told the South Korean National Assembly that North Korea had stockpiled between 2,500 and 5,000 tons of chemical weapons, and also possessed germ warfare stockpiles of anthrax and smallpox, with appropriate deployment weapons.

At that time, such unwelcome facts inevitably suggested that the survival of a Communist regime in North Korea encouraged the view that a resumption of hostilities could not yet be ruled out.

This view was strengthened by another clash between the two navies on 28 June 2002, when a South Korean frigate was sunk by North Korean gunfire, which killed four South Korean sailors and injured nineteen. Returning the fire, the South Korean Navy killed or injured thirty North Koreans.[6] This incident seemed to support the inclusion of North Korea in President Bush's widely reported "axis of evil" State of the Union speech in January 2002, along with Iraq and Iran.

Nevertheless, U.S. relations with North Korea improved dramatically when the U.S. secretary of state, Colin Powell, met Pyongyang's foreign minister, Pack Namsung for a fifteen-minute talk on the sidelines of an Asian regional security summit in Brunei on 31 July the same year.[7]

It would, of course, be going too far to conclude from this brief encounter that Korea has ceased to be a threat to international peace.

Epilogue: Is Peace Possible?

There is a tinge of political naivety in the title of this epilogue. Certainly, in the history of the world up to and including the twentieth century, there was nothing to encourage the view that world peace was possible, let alone likely. The United Nations was set up with this unlikely object in mind, and since that time, the UN itself has been involved in various conflicts, such as the Korean War, the Gulf War, and the Kosovo war.

France in particular involved itself in colonial wars designed to keep the fruit of past wars of conquest (notably in Algeria and Indochina). As for the United States, it involved itself in several wars, three of which ended in military victory, and only one in political victory as well. Moreover, although not in possession of colonies striving for independence, it decided—shortsightedly—to intervene in, or launch, several other wars, which ended in varying degrees of failure. These wars have already been described in this book, but it may be useful to summarize their outcomes here:

• The First World War. Without U.S. intervention, it is likely that the two major Western powers involved—France and Britain—would have been defeated by the powerful German forces. In effect, not only France and Britain, but also Belgium, Holland, and Italy would have become subjugated nations under German control. In such an unbearably dark event, a military and imperial Germany would have attempted, a generation or so later, to subjugate the rest of Europe, including Communist Russia; and possibly would have succeeded. (This is of course speculative but possibly useful fiction.

• The Second World War. Of all the wars considered in this book, World War II was by far the most successful, if measured (as I have done) by the ultimate political and social criteria of democracy. In particular those two major aggressive countries of the twentieth and past centuries, Germany and Japan, emerged from military defeat with encouraging signs in recent decades to support the view that they have abandoned war as an instrument of policy.

These undoubted successes may well, if only subconsciously, have encouraged the United States, with the best of intentions, to intervene in three later wars, none of which was politically successful:

• Korea, Vietnam, Algeria, and Iraq. It would be misleading, of course, to assert that the United States, as the major participant in the United Nations intervention in the Korean War, was defeated in a military sense. Certainly, the UN forces failed to wrench North Korea from the Communist regime imposed on it by Soviet and Chinese power. On the other hand, if the war was not won, in the full sense, at least it consolidated anti-communist power in South Korea, leading to another political victory for democracy.

Again, no half-success was achieved in France's protracted attempt to prevent a communist takeover of Vietnam, culminating in the humiliating defeat of Dien Bien Phu. The same was true of the protracted French attempt to preserve France's control over its North African territory of Algeria, which lasted from 1945 to 1962 and ended in failure.

No such half-success rewarded America for its own post-colonial Vietnam War. After fourteen years of intervention, under presidents Kennedy, Johnson, and Nixon, the Vietnamese Communists brought the whole of Vietnam under their control; plus the other former French colonies of Laos and Cambodia (now Kampuchea). At least it can be said in their favor that they rescued Cambodia from the horrendous crimes of Pol Pot.

Of the Soviet intervention in Afghanistan, its devastating failure was a blessing to the world at large, as it undoubtedly precipitated the collapse of the Soviet Union; which of course was not the intention of the Soviet invaders.

As for the two Gulf Wars, there is nothing good to be said about the first, and hardly anything in favor of the second. The first was an exercise in ambitious vanity by the Iraqi tyrant, Saddam Hussein; which, fortunately for the world at large, ended in failure for the invader. As for the second, the military "victory" of the UN coalition led by President George Bush, Sr. was nullified by the political failure to remove the evil dictator Hussein.

As for the United Kingdom, at least one political as well as military victory of consequence was achieved through the Malayan Emergency of 1948-1960. It is worth noting, of course, that if the British aim had been to bring Malaya and Singapore back under British colonial rule, a failure similar to the French failures in

Indochina and Algeria would have ensued. Fortunately, this was not the case, and the beneficiaries were the populations of the two ex-colonies, from the triumph of democracy over a determined Communist challenge.

One of the three "minimal wars" also described tells a less glorious story. The Suez campaign was a Franco-British absurdity, consolidating the dictatorship of Colonel Nasser in Egypt. The devastating military success of Israel in its Six-Day War did nothing to protect the Jewish state from the continuous harassment from the Palestinian minority and its Arab supporters.

The Falklands War was a success for Britain, although whether it was worth the cost of military intervention is debatable.

• India v. Pakistan, and Kosovo. The long-lasting and dangerous rivalry between India and Pakistan continues, increasingly focused on predominantly Muslim Kashmir, incorporated into predominantly Hindu India because of Nehru's sentimental attachment to the province of his ancestors.

Kosovo is a different matter: NATO's offensive against Serbia was the first in its half century to be launched without UN approval and outside the terms of reference of the Atlantic Treaty. However, it was a political success in that it led to the overthrow of the tyrannical Serbian dictator, Slobodan Milosevic, later tried on charges of war crimes in the International Court of Justice in The Hague. A major victory for democracy followed in Kosovo.

Some reflections on the methods of war are needed. Air attacks have played an increasingly important role in wars of the second half of the twentieth century; but air raids, although they may bring military victory, do not necessarily bring political victory: the theme of this book. A possible exception to this observation was America's victory over Japan in World War II, but with certain reservations. The virtual destruction of Hiroshima and Nagasaki did bring victory to the United States, but only because, for the first time in history, the American side used atomic bombs: to which the Japanese had no answer. But of course the Americans had also fought a long and costly ground war, and were well placed to take over politically when the Japanese pleaded for peace.

The same was true, on a different scale, of Britain's Royal Air Force bombings of German cities (in particular of Dresden) in the last phase of World War II. In themselves, the mass destruction of cities and killing of inhabitants, would not have brought victory, without the presence of ground forces to occupy the towns that had been destroyed.

Moving to more recent times, the concentrated American bombing of Iraq during the Second Gulf War (see chapter 19) did bring military victory to President Bush's coalition, but certainly not political victory, mainly because there was no Allied ground occupation of Iraq and especially of Baghdad.

• The Islamic terrorist offensive. On the morning of 11 September 2001, two small teams of Islamic suicide terrorists hijacked two passenger planes and crashed into the twin towers of the World Trade Center in New York. By definition, the suicide terrorists lost their lives, to nobody's regret, except possibly from their distant families. Each of the planes was a Boeing 767. The dead on the first included ninety-two passengers, and the second sixty-five, plus crew in each case. That same day, a hijacked Boeing 757 with sixty-four passengers and crew on its way to Los Angeles crashed into the Pentagon, on the outskirts of Washington. A fourth hijacked airliner crashed in Pennsylvania, without achieving its possible objectives.

When the two 110-storey towers of the World Trade Center collapsed. an estimated total of 400,000 tons of rubble, glass and dust fell into the streets of Manhattan. The dead were estimated at 6,500 (later reduced by about 55 percent).

The vicious and unpredictable attack gave a new twist to international terrorism. The motivation of the terrorists was in no sense communist; nor was the ex-Soviet Union in any way involved in it. A new threat to Western civilization had emerged: a fanatically anti-Western and especially anti-American Islamic terrorist network calling itself Al-Qaeda ("The Base") created and controlled by a merciless, Saudi-born terrorist named Osama bin Laden, operating originally from Afghanistan.

Although Islamic terrorist groups have been active for many years, they have mostly been on a small scale, whereas the Al-Qaeda network is internationally wide and backed by considerable funds. Moreover, most of the earlier terrorist networks were ideologically communist and heavily dependent on Soviet support, training, arms and equipment.*

A comparison of the former Communist terrorist network and the newly emerged Islamic one may be helpful to the reader.

• The Communist network. At its height, communism was the major threat to world peace, and by far the major source of international terrorism: that is, communist-inspired and/or communist-supported terrorism. Its hold on terrorist movements was not universal, however, for in a number of countries, nationalist rather than communist terrorism prevailed, for instance in Ireland, through the IRA (Irish Republican Army), whose aim was to incorporate British-ruled Protestant Ulster (in Northern Ireland) into the independent Irish Republic. Another example was ETA (Euzkadi ta Azkatasuma: Freedom for the Basque Homeland, that is, freedom from control by Spain). It is relevant to add, however, that both the IRA and ETA had accepted Soviet assistance.*

It should be noted, moreover, that the Soviet collapse was not necessarily followed by the collapse of Soviet-supported groups. The most interesting example of terrorist survival, in this sense—but not the only one—is the continuing activity of the FARC (Armed Revolutionary Forces of Colombia), founded in 1966 under Soviet control. There is no sign, however, of continued support from Russia since the Soviet collapse.

Broadly speaking, the Soviet bloc in Eastern Europe (USSR, Poland, Czechoslovakia, Hungary, East Germany, Bulgaria Albania, Rumania) was under Moscow's control, plus Cuba, Grenada, and Afghanistan. However, Grenada was in effect wrested from Soviet control by the U.S. occupation in 1983; while Afghanistan regained its independence with the Soviet defeat and withdrawal in 1988 (when it passed under the control of the extremist Islamic movement called the Taliban). East Germany merged with the (West) German Federal Republic in 1989-90.

Communist governments were of course also in power in a number of countries no longer controlled by Moscow. These included (and still include) China, Vietnam, Mongolia, Cambodia, Laos, and Cuba, The four Asian communist countries had long ceased to be controlled by Moscow. Cuba remained under strong Moscow influence until the Soviet collapse. At the height of Moscow's imperialist expansion, the total number of Communist-controlled countries was eighteen. At the time of the Soviet collapse, the figure had been

---

*My first book on international terrorism, *The Rebels*, appeared in 1960 and became "required reading" in NATO and SEATO (South-East Asia Treaty Organization). The systematic study of terrorism was a main function of the London Institute for the Study of Conflict (ISC) which I established in 1970 and directed for nine years. By 1973, terrorist centers had multiplied, and my analysis of what I termed *"Transnational Terrorism"* was the title of my introductory article in the ISC's Annual of Power and Conflict for that year.

reduced to eleven. It should be added that Yugoslavia, which in fact had autonomous parties for each of the six component republics, plus the League of Communists of Yugoslavia as the sole governing party of the "Socialist Federal Republic of Yugoslavia," had long ceased to be an instrument of Soviet foreign policy.

At the height of Soviet power, there were 128 Communist parties worldwide, a majority of them, but not all, controlled from Moscow.[1] In regional alphabetical order, the total was as follows:

Africa	12
The Americas	34
Asis/Pacific	26
Eastern Europe and USSR	9
Middle East	17
Western Europe	30
Total	128

In the 1960s, at the height of the protracted split between Russia and China, the Chinese Communists launched a number of subversive activities in Africa, most of which failed. They also spread their influence, more successfully, in South-East Asia.[2]

• The Islamic Challenge. In comparison with communism, the spread of Islam, in terms of Islamic States is considerably wider. However, if a list of Islamic states is limited to those involved, or likely to be involved, in international terrorism, the number is considerably smaller. A provisional—and tentative—list follows:

Afghanistan; a Communist danger for some years, but much in the news on and after the 11 September incidents, initially as the place of refuge of Osama bin Laden, and with the extremist Muslim Taliban in power. However, the Taliban regime was ousted as a result of U.S. (and U.K.) military intervention (2001–020), and Bin Laden vanished.

Algeria: Islam is the State religion, but under a military regime.

Iran: Since 1 April 1979, after the expulsion of the Shah, Iran has been under strictly orthodox Islamic rule and therefore constitutes a potential international danger.

Iraq: military dictatorship under Saddam Hussein. A Shia majority but ruled by a Sunni minority. Reported supporter of anti-Western terrorism.

Libya: In effect a dictatorship under Colonel Qaddafi, personally involved in various terrorist operations against the West Pakistan. A major Islamic country, but helpful to the U.S. and its allies after the 11 September terrorist outrages.

Saudi Arabia: an absolutist Islamic regime whose capital, Mecca, is in effect the centre of the Muslim world. As an oil-rich country heavily dependent on Western economic links, unlikely to involve itself directly in international terrorism beyond, on occasion, providing hospitality for terrorists.

Syrian Arab Republic: a possible ally in Islamic terrorism.[3]

An ideological comparison of terrorist motivation for Communists and Muslims is worth making. The two are of course totally incompatible. Communist regimes justified terrorism on ideological grounds; Islamic supporters of terrorism, on religious grounds. In Communist regimes, from the first one under Lenin, violence and killing were necessary to the seizure and especially the maintenance of power. Marxist doctrine, as interpreted by Lenin as the first Marxist revolutionary leader to seize power, justified the destruction of "capitalism" as the only way to introduce the material paradise of communism.

It followed, throughout the seventy-four years of the Soviet regime, that terrorist gangs in "capitalist" countries must be supported (with money, arms, ammunition and training) as a recognized means of destroying "capitalist" (i.e., democratic) regimes and replacing them with communist ones until the whole world became communist. In theory (though remote from practice) the "proletariat" would be in power, permanently. In China, the communist leader, Mao Zedong, viewed terrorism as the necessary prelude to guerrilla warfare, enabling the communist guerrillas to seize power. Once power had been seized, terror (as distinct from terrorism) would maintain that power indefinitely.

Despite the strikingly different Islamic justification of terrorism, the outcome would in some respects be remarkably similar. Islamic terrorists kill in the name of Allah (God) and of his Prophet, Mohammed. The unbelievers (Infidels) may therefore be killed with the approval of Allah and his Prophet. Because of this "approval," accepted as such by the faithful, they have everything to gain, and nothing to lose, by killing the infidels, whether Jews (in Israel) or Christians, defined sweepingly as the inhabitants of non-Islamic countries. The deliberate suicide of terrorists is justified on the argument that they are furthering the aims of Islam; for this, they will be rewarded by ascent to Paradise.

Clearly, as demonstrated by the events of 11 September in the "infidel" city of New York, the self-use of suicide terrorists substantially increases the efficacy of terrorism and brings an immediate (if imaginary) reward. Unless and until the religious leaders of Islamic countries personally and publicly condemn suicide bombings, the myth of paradisial rewards will continue to flourish.

• The anti-terrorist offensive. America's much-criticized new leader, President George Bush, son of his president father of the same names, rose to the threat with impressive speed and determination. In a televised speech to the nation on the evening of 11 September, he described the attacks as "more than acts of terror. They are acts of war." In speeches over the next few days, he described the war against terrorism as "a crusade," and (on 20 September) he said: "The war on terrorism will not end until every terrorist group of global reach has been found, stopped and defeated."

The terrorist attacks were condemned by Prime Minister Tony Blair of Britain and President Vladimir Putin of Russia. The Palestinian leader Yasser Arafat, perhaps surprisingly, denounced the attacks as "a terrible act"; even President Jiang Zemin of Communist China condemned them. Not surprisingly, President Saddam Hussein of Iraq refrained from any such criticism, remarking on the day of the attacks that the U.S. was "reaping the thorns of its foreign policy." Interestingly, President Mohammad Khatami of Iran condemned the attacks, but his religious superior, Ayatollah Khamenei, ruled out any form of Iranian cooperation with the U.S. (as had been hinted by Khatami).

Encouragingly, NATO officials meeting in Brussels on 12 September 2001 invoked Article 5 of the NATO Charter stating that an attack on one member state was considered an attack on all 19 members. Public support also came from Germany, France, India, and Pakistan, among others that included the former Soviet Islamic republics of Kazakhstan, Turkmenistan, Tajikistan, and Uzbekistan. Pakistan's offer of help, through president general Pervaiz Musharraf, was of special value because of Pakistan's border with Afghanistan.

In effect, the Western war against Al Qaeda and its political support from the Islamic extremist Taliban government of Afghanistan, was conducted by U.S. air and ground forces, with ground support from British special forces. Within weeks (by early December 2001), the Taliban forces had been defeated: the capital, Kabul, had been captured on 12 and 13 November, and the Pashtun tribal leader, Hamid Karzai—a moderate Muslim with a fluent command of English—had taken over from the Taliban leader, Mullah Mohammed Omar. Among the rejoicers was the feminine population who could now remove their impenetrable veils and show their faces to the world.

The Taliban defeat was the good news. The bad news was that all attempts to find and detain (or kill) the terrorist leader Osama Bin Laden, had failed. Nobody on the Western side knew where he had taken refuge; or indeed whether he was still alive. So the struggle went on.

So, is peace possible? The simple answer is Yes. After all, in past centuries, West European nations were often at war with each other, not least England versus France on numerous occasions. Yet now, in this twenty-first century, major European wars seem very unlikely, despite Russia's second attempt to crush the Chechens, and the Western intervention in Kosovo.

But it would be foolishly optimistic to suppose that permanent peace has come to the near East, South-East Asia, Africa, or Latin America.

Political victory, followed by peace, is certainly possible, but only in the event of the spread of democracy worldwide (in some future generation).

Meanwhile, if ever a democracy finds itself facing military hostility, the real lesson of the selected wars analyzed in this book is for political leaders, at all times, to maintain their authority over their military forces. Only thus, can political victory be made available.

Notes

Chapter 1

1. Sources consulted: *An Encyclopedia of World History*, compiled and edited by William L.Langer (Houghton Mifflin, Boston, 1968). Henceforth, *World History*. *The New Encyclopaedia Britannica* (University of Chicago, 1975 ed.). Henceforth, *Britannica*. And the *Grand Larousse Illustré* (Librairie Larousse, Paris, 1968). Henceforth, *Larousse*.
2. Ibid..
3, Brian Crozier, *The Rise and Fall of the Soviet Empire* (Prima, Rocklin, California, 1999-2000), p.8. Henceforth. BC, *Empire*. For further details, Dmitri Volkogonov, *Lenin: Life and Legacy* (trans.Harold Shukman, HarperCollins, London, 1995, p.110); and Richard Pipes, *The Russian Revolution, 1899-1919* (Alfred A.Knopf, New York, 1990), Ch.11, p.411.
4. These passages, and indeed this chapter as a whole, owe much to the splendidly detailed and highly readable book, *Peacemakers: The Paris Conference of 1919 and its attempt to End War*, by Margaret Macmillan (John Murray, London, 2001). Henceforth, *Peacemakers*.
5. See 1 above.

Chapter 2

1. For events leading to the rise of Hitler, the best source available is William L. Shirer's classic *The Rise and Fall of the Third Reich* (Secker and Warburg, London, 1963). See, in particular, the first six chapters. Henceforth, Shirer.

Chapter 3

1. Shirer, pp.193-185.
2. Shirer, p.189.
3. Brian Crozier, *De Gaulle* (HarperCollins, London, 1973, with introduction by Richard Nixon, The Easton Press, Norwalk, CT), p.605.
4. Brian Crozier, *Free Agent: The Unseen War 1941-1991*, pp.97-101, 131, 135, 157, 159, 191-3, 209, 214-17, 241. Henceforth, BC, *Agent*.

Chapter 4

1. *Pravda*, 3 September 1945, quoted by David Rees, *The Soviet Seizure of the Kuriles* (Praeger, New York, 1984, p.82). For further details, see BC, *The Rise and Fall of the Soviet Empire* (Prima Publishing, Roseville, California, 1999), p.87. Henceforth, BC, *Empire*.
2. David Rees. *Korea: The Limited War* (Macmillan, London, 1964), pp.214 et seq. Henceforth, Rees, *Korea*.

Chapter 5

1.   For a full, though favorably biased, biography, see Jean Lacouture, *Ho Chi Minh* (Seuil, Paris, 1967); for a detailed account of his early career, Brian Crozier, *The Rebels* (Chatto and Windus, London, 1960). Henceforth, BC, *Rebels*.
2.   Quoted in BC, *South-East Asia in Turmoil* (Penguin, 3rd ed., 1968), p.47. Henceforth, BC, *Turmoil*.
3.   Donald Lancaster, *The Emancipation of French Indo-China* (Oxford University Press, 1961), p. 218. Henceforth, Lancaster.
4.   Some months after General de Lattre's death, I arrived in Hanoi as a Reuters news agency correspondent, where I had a long conversation with the French commander in Tonking, General de Linarès. In negative praise for the French forces on their withdrawal from Hoa Binh, he said: "What I like about [it] was that it was a manoeuvre of perfect military orthodoxy." On General Giap: "I wish I had him on my side" (BC, *Turmoil*, p.56).
5.   For a detailed account of the 1954 Geneva conference, see Lancaster, op. cit., ch.XVII. For a concise summary, BC, *Empire*, pp. 128-129.

Chapter 6

1.   BC, *Rebels*, p.20.
2.   Ibid., p.19.
3.   Ibid., p.21.
4.   Ibid., p.24.
5.   BC, *The Masters of Power* (Eyre & Spottiswoode, London, 1969), p.20. Henceforth, BC, *Masters*.
6.   These passages draw heavily on the *World Encyclopedia of Political Systems and Parties*, ed. George E. Delury (Facts on File, New York, 1983)., pp.12-14. Henceforth, *Systems*. George E. Delury (Longman, pp.12-14).
7.   For further details, see Michel Gurfinkiel in *Valeurs Actuelles*, the Paris news magazine, 4 May 2001.
8.   Quoted in *Valeurs Actuelles*, 4 May 2001.

Chapter 7

1.   The most authoritative account of the Calcutta Conference of Democratic Youth is in J.H. Brimmell's important book, *Communism in South East Asia* (Oxford University Press, 1959). See also BC, *Empire*, pp. 130-131.
2.   This quotation is from p.17 of Harry Miller's excellent book, *Jungle War in Malaya: The Campaign against Communism, 1948-60* (Arthur Barker, London, 1972). The author was a senior colleague of mine on the *Straits Times* in Singapore in 1952-53. Henceforth, Miller.
3.   See John Cloake, *Templer: Tiger of Malaya* (Harrap, London, 1985), pp.196 et seq. The best and fullest biography of General Templer The author, John Cloake, a career diplomat who was also stationed in Singapore during my time there. By the time this splendid volume appeared, his subject was Field Marshal Sir Gerald Templer, a worthy recompense for a brilliant career. Henceforth, Cloake.
4.   See also Miller, ch.7, "Briggs and his Master-Plan," pp. 69 et seq.; and Sir Robert .Thompson, "Emergency in Malaya: The New Village Strategy" in *War and Peace: An Analysis of Warfare since 1945* (Orbis, London, 1981, pp. 86-87). Henceforth, *Warfare*. The most detailed account of the Briggs plan and the New Villages is in

Anthony Short, *The Communist Insurrection in Malaya, 1948-1960* (Frederick Muller Ltd, 1975); in particular in chapters 9 and 15.
5. Quoted in Miller (see 3 above), p.846. For a detailed account of Strategic Hamlets, see Sir Robert Thompson, *Defeating Communist Insurgency* (Chatto & Windus, London, 1966), ch.11. Henceforth, Thompson, *Insurgency*. Having been posted to Vietnam. at the invitation of the U.S. government, to provide guidance on tackling the Communist guerrillas, Thompson in fact devotes considerably more space in this chapter to his work on similar lines in Vietnam.
7. Thompson, *Insurgency*, pp.50-62.
8. Miller, pp.117 et seq.
9. The casualty and other figures quoted are in Miller, pp.181-2.

Chapter 8

1. Brian Crozier, Drew Middleton and Jeremy Murray-Brown, *This War Called Peace* (Sherwood Press, London, 1984). pp.182 et seq. Henceforth, *War/Peace*.
2. Thompson in *Warfare* (see ch.7, note 4 above).
3. Jonathan Aitken, *Nixon: A Life* (Weidenfeld & Nicolson, London, 1993), p.386. An important study of American politics during Nixon's life. Henceforth, Aitken.
4. Aitken, pp.384-385.
5. Richard Nixon, *The Real War* (Warner Books Inc., 1980), p.114.
6. Ibid., p.115.
7. *Times* (London) 23 April 1975.
8. Thompson, *Insurgency*, p.25. See also BC, *Power* (see ch.6, n.5 above), pp. 242-246.

Chapter 9

Of the many sources consulted, the most helpful were: *Chronicle of the 20th Century* (Longman, London, 1988); as always, Keesing's, *Record of World Events*, together with the *Encyclopedia of World History*.

Chapter 10

1. *This War Called Peace*, op. cit., p.197.
2. BC, *Empire*, p.302

Chapter 11

1. Michael Orr, "The Soviet Intervention," in *War and Peace*, p.288.
2. BC, *Empire*, p.308.
3. Orr, "Soviet Intervention," 308.
4. Ibid.
5. Ibid.
6. In Alex Alexiev's well informed article in *Global Affairs* (Winter 1988). Henceforth, Alexiev.

Chapter 12

1. Alexiev (see note 6 in ch.11).
2. *Washington Post*, 13 January 1985.

3.   Senator Malcolm Waller, "US Covert Action," *Strategic Review*, Winter 1984, p.12.
4.   Aaron Karp, "Blowpipes and Stingers in Afghanistan: One Year Later": *Armed Forces Journal International*, September 1987.   .

Chapter 13

1.   This chapter is factually based entirely on the explicit and reliable coverage of Keesing's *Record of World Events, 1990-96*. Henceforth, Keesing.

Chapter 14

1   For a full account of the Grenada incident, see BC, *Empire*, ch.38, pp.370 et seq..
2.   For a full analysis of Gorbachev's policies, see BC, *The Gorbachev Phenomenon* (Claridge, London, 1990). Henceforth, BC, *Phenomenon*.
3.   Quoted (with a minor correction) from p.392 of BC, *Empire*.
4.   In effect, a summary of Lenin's stated aim.
5.   Quoted from Robert Bozzo, "Gorbachev and the Nationalities Problem," in *Global Affairs* (Summer/Fall 1990); a succinct analysis of an important, but largely neglected aspect of Gorbachev's policies.
6.   The ensuing paragraphs incorporate corresponding passages in my book, *Phenomenon*, pp.71 et seq.
7.   As reported in the left-of-center London daily, the *Guardian* of 5 December 1989 under the headline "Russians condemn invasion of Prague."
8.   Markus Wolf (with Anne McElvoy), *Man Without a Face: the Memoirs of a Spymaster* (London, Jonathan Cape, 1997), pp 324-325. See also David Pryce-Jones, *The War that Never Was* (London, Weidenfeld & Nicolson, 1995), pp. 239 et seq.

Chapter 15

1.   This account draws heavily from my book, *Masters* (see ch.6, n. 5.), in Part 3, 1, "A Suez Success Story," a deliberate mixture of fact and fantasy.
2.   Hugh Thomas, *The Suez Affair*, p.70. An excellent account. (Weidenfeld & Nicolson, London, 1967).
3.   For a full and authoritative account of the Anglo-French operation, and the Israeli invasion of Egypt, see H. P. Willmott, *The Suez Fiasco in Warfare*.
4.   Masters, pp.119 et seq.

Chapter 16

1.   H. P. Willmott, at the time of his relevant chapter in *Warfare* (1981) was a senior lecturer at Sandhurst. Henceforth , Willmott.
2.   Willmott. p.162.
3.   These paragraphs about Arafat are adapted from my lengthy profile of the man ("Arafat: From Gun-Runner to Diplomat") in the (now defunct) news magazine *Now*! of 3 April 1981.
       Also consulted:
       Jonathan Dimbleby, *The Palestinians*: a well-written and profusely illustrated , pro-Palestinian analysis by a well-known British television presenter; (Quarter Books London,1979); and Joan Peters, *From Time Immemorial: The Origins of the Arab-Jewish Conflict over Palestine* (Michael Joseph, London, 1984): a massive, basically pro-Jewish but remarkably objective historical account of the historical roots

of the Arab-Jewish conflict. As with many chapters of this book, much research was guided by *Britannica, World History, Everyman's Encyclopaedia*; and the indispensable Keesing. To these should be added the London-based *Institute for the Study of Conflict's Annual of Power and Conflict*, indispensable for the study of terrorism in the 1970s.

Chapter 17

1.   *The Falklands War: The Full Story by the Sunday Times Insight Team* (Sphere Books, London 1982), henceforth, *Insight Report*.) A remarkably full and objective account, upon which I draw with gratitude.
2.   *Insight Report*, p.25.
3.   Ibid., p.26.
4,   John Nott, *Here Today Gone Tomorrow: Recollections of an Errant Politician* (Politico's, London, 2002), p.249. Another full, but personalized, account of the Falklands War.
5.   *Insight Report*, p.263.

Chapter 18

1.   For a fuller account of the Shah of Iran's removal, see BC, *Free Agent* (Harper Collins, London, 1993), pp.170 et seq.
2.   For these and other relevant details, I am indebted to Michael Orr, lecturer at Sandhurst and author of the chapter on the Iran-Iraq war in the authoritative collective study, *Warfare*. Henceforth, Orr.
3.   Orr, p.297.

Chapter 19

1   For a full account, see *World Encyclopedia of Political Systems* (Facts on File, New York, 1983; ed. George E. Delury), pp.493 et seq., Henceforth, Systems. "Republic of Iraq" (Al-Jumhuriyya al-'Iraquiyya), by Revva Simon. See also Brian Crozier, *National Review*, New York, 18 March 1991 (henceforth, *NR*).
2.   *NR*, 18 March 1991.
3.   *NR*, 13 May 1991.
4.   *Newsweek*, 1 December 1997.
5.   *American Legion*, August 1995.
6.   At the time of the Allied victory, in February-March 1991, the controversy over whether or not the Iraqi dictator should be assassinated was raging. There were widespread reports to the effect that General Schwarzkopf was in favor and General Powell against. However, in Schwarzkopf's autobiography, *It Doesn't Take a Hero* (written with Peter Petre), the general does not admit, let alone boast of, having advocated the assassination or kidnapping, or bringing to trial of the Iraqi dictator. The nearest he comes to it is in a brief passage on p.400: "The next question I inevitably get is a follow-up one: Since Saddam is still alive and in control in Iraq, wasn't the whole war fought for nothing? I will confess that emotionally, I, like so many others, would have liked to see Saddam Hussein brought to some form of justice. He may still be...."
     Understandably, this question had aroused considerable discussion and controversy. On 2 March 1998, for instance, the Sub-Committee on Near Eastern and South Asian Affairs of the Committee on Foreign Relations of the U.S. Senate held a hearing on the subject. At that time, the Committee included the well-known names

of Senators Jesse Helms (North Carolina), the chairman; and Richard G. Lugar (Indiana). The witnesses included an opponent of Saddam, Ahmed Chalabi, president of the Iraqi National Congress, at that time based in London; and R. James Woolsey, a former Director of the U.S. Central Intelligence Agency (CIA).

In an opening statement, Senator Sam Brownback (Kansas) referred to Hussein as "the root cause" of the continuing problem of Iraq. The first Statement was by Chalabi (see above), who said, inter alia: "Saddam is a mass murderer who is personally responsible for the genocidal slaughter of at least 200,000 Iraqi Kurds, 250,000 other Iraqis . . . and tens of thousands of Iraqi dissidents in the last ten years . . . Now that Saddam again threatens not only the Iraqi people but the region and the world, the Iraqi people ask you to give us the tools and let us finish the job."

The former CIA chief, Woolsey, took an emollient line: "I do not believe that we should attempt to assassinate Saddam or even arrange a coup against him....I believe that as the world's dominant political, economic and, in many ways, cultural power, the United States should not tell the world now that we support assassination as a tool of statecraft."

Also consulted: Saïd K.Aburish, Saddam Hussein: *The Politics of Revenge.*
Andrew and Patrick Cockburn, *Out of the Ashes: The Resurrection of Saddam Hussein.*

Chapter 20

1.   Peter Duignan, *NATO: Its Past, Present, and Future* (Hoover Institution Press, Stanford University, Stanford, CA, 2000): Part IV, "The Yugoslav Imbroglio," pp. 85 et seq.
2.   This theme is extensively and intelligently discussed by Peter Duignan, *NATO*. See also "The Future of NATO," *Economist*, 4-10 May 2002.
      Other sources
      Noel Malcolm, *Kosovo: A Short History* (Macmillan, London, 1996). An erudite and stylishly written volume, which, at 492 pages, contradicts the term "short" in its title.
      M.A. Smith, *Kosovo: Background and Chronology* (Conflict Studies Research Centre, Royal Military Academy, Sandhurst, April 1999). The comprehensively detailed chronology deals with events in the crucial period of January-April 1999.
      Hoover Digest (Hoover Institution), nos. 1 and 4, 1999.

Chapter 21

1.   The death tolls attributed to the Soviet and Chinese Communist regimes are cited in the best-selling book *Le Livre Noir Du Communisme: Crimes, Terreur, Répression* (Robert Laffont, Paris, 1997), figures quoted on p.14. The authors: Stéphane Courtois, Nicolas Werth, Jean-Louis Panné, Andrzej Paczkowski, Karel Bartozek, Jean-Louis Margolin. Henceforth, *Livre Noir*.
2.   This short account of Mao's early life is adapted from Eric Chou's biography of the Communist leader, *Mao Tsetung: The Man and the Myth* (Cassell, London, 1982), specifically from the author's detailed chronology, pp.253-260.
3.   For a fuller account of the Mao-Stalin meetings, see BC, *Empire,* ch. 14, with verbatim accounts in the appendices.
4.   See Stuart Schram, *Mao Tse-tung* (Pelican, London 1966), pp.254-257; and Harold M. Vinacke, *Far Eastern Politics in the Postwar Period* (London, 1956), pp.170-71.

5. For further details, see Richard Evans, *Deng Xiaoping and the Making of Modern China* (Hamish Hamilton, London, 1991), p.153. Henceforth, *Deng*. For further details, see BC , *Since Stalin: An Assessment of Communist Power* (Hamish Hamilton, 1970), part II, ch.2, "China on the Rampage."
6. *Deng*, p.263.

Chapter 22

1. These details are drawn from the author's full biography of Chiang Kai-shek, *The Man Who Lost China* (Angus & Robertson, London, 1977, original edition, Charles Scribner's Sons, New York, 1976), pp. 327 et seq. Henceforth: BC, *Chiang*.
2. BC, *Chiang*, p.328.
3. In greater detail, these facts and figures first appeared in the originally unsigned *Economist* articles, in the issues dated 2 and 9 February 1957. They were reproduced, with permission from the editor of the *Economist* and with attribution to the author, in a Pamphlet published in March 1957 by the Friends of Free China Association, in London, under the title "I was wrong about Free China." Also reproduced was a talk on the same theme which I gave on the BBC, and which the journal the *Listener* reproduced on 21 February that year.

Chapter 23

1. This summary of events draws heavily on p.4 of David Rees's key book, *Korea* (see ch.4, n. 2: p.4 et seq. Henceforth, Rees.
2. See the three authors' account of the *Cold War, This War Called Peace*, by Brian Crozier, Drew Middleton and Jeremy Murray-Brown (Sherwood Press, London, 1984), ch. 4, "Hot War in Asia," pp. 88 et seq.
3. For full Soviet documentation of the crucial meeting between Stalin and Kim Il Sung, see appendix E in BC, *Empire*, pp. 579 et seq.
4. Rees, chs. 1 and 2.
5. Ibid.
6. For a fuller account, see the *Economist*, 6 July 2002, p.61.
7. For a full account of the Powell-Pack Namsun meeting, see the *Daily Telegraph* (London, 1 August 2002); and the *Economist*, 3 August 2002.

Epilogue

1. For fuller details, see *1988 Yearbook on International Communist Affairs*, ed. Richard F. Staar (Hoover Institution Press, Stanford, California).
2. For further details of the Sino-Soviet subversive rivalry, see Brian Crozier, *The Struggle for the Third World* (Background Books, The Bodley Head, London, 1966); and especially *Empire*, cch.24, "The Rival Suns."
3. Details drawn partly from the two-volume *Systems*.
4. For a full coverage of all Muslim (as well as non-Muslim) states, see *Systems*.
5. Quoted from the Koran, Sura XLVII, in the translation from the Arabic by the Rev. J. M. Rodwell (Everyman's Library, Dent & Sons, London and New York, 1929)., p.382. The translator adds the following footnote: "This Sura was revealed at a period after the victory at Bedr, when there was still some hesitation on the part of Muhammad's followers to take decided steps for securing their position." As with all the Suras, this one was preceded by the ritual words: In the Name of God, the Compassionate, the Merciful. Was the Almighty less compassionate and merciful where infidels were concerned? This is presumably the view taken by Osama Bin Laden and by the suicide bombers of 11 September 2001.

Index